STRAIGHT UP

JOSEPH J. ROMM

STRAIGHT UP

America's Fiercest Climate Blogger
Takes on the Status Quo Media, Politicians,
and Clean Energy Solutions

JOSEPH J. ROMM

Washington | Covelo | London

Island Press, Suite 300, 1718 Connecticut Ave., NW, Washington, DC 20009

Library of Congress Cataloging-in-Publication Data

Romm, Joseph J.
Straight up : America's fiercest climate blogger takes on status quo media, politicians, and clean energy solutions / Joseph J. Romm.
p. cm.
Includes bibliographical references and index.
ISBN-13: 978-1-59726-716-8 (pbk. : alk. paper)
ISBN-10: 1-59726-716-3 (pbk. : alk. paper) 1. Climatic changes—United States. 2. Global warming—Press coverage—United States. 3. Communication in science—Political aspects—United States. 4. Scientists—Psychology. I. Title.
QC903.2.U6R66 2010
551.6--dc22

2009051086

Printed using Copperplate Gothic and Minion

Text design and typesetting by Joan Wolbier

Printed on recycled, acid-free paper

Manufactured in the United States of America

10 9 8 7 6 5 4 3 2 1

Keywords: clean energy, climate change, ClimateProgress.org, climate progress, Center for American Progress, Time Magazine Heroes of the Environment, renewable energy, right-wing disinformation, IPCC (Intergovernmental Panel on Climate Change), climate science, climate denier, concentrated solar power, energy efficiency, atmospheric carbon, emissions scenarios, energy policy, mitigation, David Broder, media accountability

To Antonia

CONTENTS

INTRODUCTION

Why I Blog

From a very early age, perhaps the age of five or six, I knew that when I grew up I should be a writer. Between the ages of about seventeen and twenty-four I tried to abandon this idea, but I did so with the consciousness that I was outraging my true nature and that sooner or later I should have to settle down and write books. . . .

I knew that I had a facility with words and a power of facing unpleasant facts. . . . —George Orwell, "Why I Write"

I JOINED THE NEW MEDIA BECAUSE THE OLD MEDIA HAVE FAILED US. They have utterly failed to force us to face unpleasant facts.

What I have learned most from the success of my blog, from the rapid growth in readers and comments, along with the increasing number of Web sites that link to or reprint my posts, is that there is in fact a great hunger out there for the bluntest possible talk. It is a hunger to learn the truth about the dire nature of our energy and climate situation, about the grave threat to our children and future generations, about the vast but still achievable scale of the solutions, about the forces in politics and media that impede action—a hunger to face unpleasant facts head on.

Unlike Orwell, I knew from a very early age, certainly by the age of five or six, that I would be a physicist, like my uncle, and I announced that proudly to all who asked.

I knew I did *not* want to be a professional writer once I saw how hopeless it was to make a living that way. My father was the editor of a small newspaper (circulation under 10,000) that he turned into a medium-sized newspaper (70,000) but was paid dirt, even though he managed the equivalent of a large manufacturing enterprise—while simultaneously writing three editorials a day—that in any other industry would pay five times as much. My mother pursued freelance writing for many years, an even more difficult way to earn a living.

Why share this? Orwell, who shares far, far more in his many brilliant essays, argues in "Why I Write":

> I give all this background information because I do not think one can assess a writer's motives without knowing something of his early development. His subject matter will be determined by the age he lives in—at least this is true in tumultuous, revolutionary ages like our own—but before he ever begins to write he will have acquired an emotional attitude from which he will never completely escape. It is his job, no doubt, to discipline his temperament and avoid getting stuck at some immature stage, in some perverse mood; but if he escapes from his early influences altogether, he will have killed his impulse to write.

And no, I'm not operating under the misimpression that my writing can be compared with Orwell's. I know of no essayist today who comes close to matching his skill in writing. On top of that, bloggers simply lack the time necessary for consistently first-rate efforts. I've written some 2 million words since launching my blog in 2006. Perfection isn't an option.

Orwell does, however, have the soul of a blogger. He has a brutal honesty that puts even the best modern memoirists to shame. And he confronts the toughest of truths, which I think is perhaps the primary quality I aspire to at ClimateProgress.org, a quality captured in the label that *Rolling Stone* gave me, "America's fiercest climate-change activist-blogger." Orwell asserts, "Putting aside the need to earn a living, I think there are four great motives for writing, at any rate for writing prose."

I see more than four great motives to blog, at least for me. But let's start with Orwell's:

> *(i) Sheer egoism.* Desire to seem clever, to be talked about, to be remembered after death. . . .

Inarguable. At least Orwell notes that "Serious writers, I should say, are on the whole more vain and self-centered than journalists." I make no pretensions to being a "serious" writer. I'm not certain that bloggers are journalists. I think we are, however, journal-ists. What is a (web) log if not a journal?

> *(ii) Aesthetic enthusiasm.* Perception of beauty in the external world, or, on the other hand, in words and their right arrangement.

I dictate all of my blog posts directly onto the PC. For me the sound of a good phrase, the pleasure of a headline that works, is immense. Sometimes everything comes together, as in perhaps my best headline, the one *Time* magazine singled out in naming ClimateProgress.org a favorite environmental website: "Debate over. Further delay fatal. Action not costly," which is the first post in chapter 2.

> *(iii) Historical impulse.* Desire to see things as they are, to find out true facts and store them up for the use of posterity.

Even more so with a blog. In the event that we don't avert catastrophic global warming, I do hope that the reporting and analysis in this blog, which evolves over time, will be of use to those trying to understand just how it is that, as Elizabeth Kolbert put it, "a technologically advanced society could choose, in essence, to destroy itself." It will be a great source of bafflement to future generations, and I suspect that, as they suffer through the misery and grief caused by our myopia and greed, a literature will emerge aimed at trying to understand what went wrong, how we did this to ourselves. Perhaps ClimateProgress.org will help.

> *(iv) Political purpose.* Using the word "political" in the widest possible sense. Desire to push the world in a certain direction, to alter other peoples' idea of the kind of society that they should strive after. Once again, no book is genuinely free from political bias. The opinion that art should have nothing to do with politics is itself a political attitude. . . .

Orwell goes on to say of himself (emphasis added),

> By nature—taking your "nature" to be the state you have attained when you are first adult—I am a person in whom the first three motives would outweigh the fourth. In a peaceful age I might have written ornate or merely descriptive books, and might have remained almost unaware of my political loyalties. *As it is I have been forced into becoming a sort of pamphleteer.*

His always careful word choice is telling. The Wikipedia entry on "pamphleteer" asserts, "Today a pamphleteer might communicate his missives by way of weblog."

> I write it because there is some lie that I want to expose, some fact to which I want to draw attention, and my initial concern is to get a hearing. But I could not do the work of writing a book, or even a long magazine article, if it were not also an aesthetic experience. . . . The job is to reconcile my ingrained likes and dislikes with the essentially public, non-individual activities that this age forces on all of us.

I couldn't dream of saying it better than that.

> And looking back through my work, I see that it is invariably where I lacked a political purpose that I wrote lifeless books and was betrayed into purple passages, sentences without meaning, decorative adjectives, and humbug generally.

I also blog for at least two other reasons.

Peace of mind. I would be unimaginably frustrated and depressed if I didn't have a way of contributing to the task of saving a livable climate, a way of responding in real time to the general humbug and the sentences without meaning and the purple passages of those who wittingly or unwittingly are spreading disinformation aimed at delaying action on climate change. I hope the comments section on the blog serves as a similar outlet for readers.

Personal growth. The act of trying to explain the science and the solutions and the politics to a broader audience forces me think hard about what I'm really saying, about what I really know and don't know. The rapid feedback and global nature of the blogosphere mean that I get to test my ideas against people who are exceedingly knowledgeable and articulate. Through this blog I have interacted with people from every walk of life, with widely different worldviews, from many continents, whom I never would have otherwise known. And all from the basement of my home, occasionally with my daughter on my lap.

It boggles the mind that I have a profession that did not exist even a decade ago, but that is, in many respects, precisely what my father did, precisely what I never expected to do.

I first became interested in global warming in the mid-1980s, while studying for my physics PhD at the Massachusetts Institute of Technology and researching my thesis on oceanography at the Scripps Institution of Oceanography in California. I was privileged to work with Dr. Walter Munk, one of the world's top ocean scientists, on advanced acoustic techniques for monitoring temperature changes in the Greenland Sea.

A few years later, as Special Assistant for International Security to Peter Goldmark, President of the Rockefeller Foundation, I found myself listening to some of the nation's top policy and security experts. Even a generation ago, they knew the gravest threats that would face us today. They convinced me that global warming was the most serious long-term, *preventable* threat to the health and well-being of our nation and the world.

In the mid-1990s, I served for five years in the U.S. Department of Energy. As an acting assistant secretary, I helped develop a climate technology strategy for the nation. Working with leading scientists and engineers at our national laboratories, I came to understand that the technology for reducing our emissions was already at hand and at a far lower cost than was widely understood—if we had smart government policies to drive those technologies into the marketplace, policies that included putting a price on carbon dioxide pollution. More recently, I have worked with some of the nation's leading corporations, helping them to make greenhouse gas reductions and commitment plans that also handsomely boost their profits.

After my brother lost his Mississippi home due to Hurricane Katrina, I started interviewing climate experts for what turned into my book *Hell and High Water*. Our top climate scientists impressed upon me the fact that the climate situation is far more dire than I had realized, far more dire than 99 percent of opinion makers and politicians understand.

I decided I would not pull any punches—I would get "political" as Orwell defined the term. I joined the Center for American Progress in 2006 because it had become the cutting-edge think tank for both policy and communications on progressive issues. I began part time, posting on my blog once a day. As readership grew and ClimateProgress.org became a leading voice on energy and climate issues, I began posting more. Now I'm a full-time blogger, writing several times a day and also featuring guest posts from some of the best writers and thinkers on the subject.

A key goal of my blog today is to save my readers time. There is far too much information on climate science, clean energy solutions, and global warming politics for anyone to keep up with. And the status quo media puts out too much analysis, most of it quite bad. And yet everyone needs to follow this issue, needs to have an informed opinion on the most important issue of the decade and the century.

And that's the goal of this book, too, to cut through the crap and focus on what's important. I have now written more than 4,000 posts. This book represents what I (and my readers) think is the best 1 percent, plus a few columns that were first published in *Salon*. I have made some minor edits in the posts. Blogging is by nature very repetitious, since in any month, a significant minority of visitors are first-timers. I have tried to edit out some of the repeated content. At the same time, some of the words and phrases may only be clear to someone who is a regular reader of my blog, so I have changed some words for clarity and consistency. Also, in a blog post, I often link to a previous post rather than breaking the flow of the argument to explain a detail. That won't work in a book. So I have deleted most of those references and explained the key point in a few sentences where necessary.

This will be my first book without an extensive set of notes. Citations allow readers to go to original sources to check the accuracy of what is written. Since all of the pieces in the book are online at ClimateProgress.org with direct hyperlinks to those original sources, it seemed redundant to reprint them here in a less useful form.

The use of some numbers to discuss energy and climate issues is both inescapable and desirable. In a blog, I can link directly to original sources for those who want more. That said, by 2020 at the latest, the key numbers—atmospheric concentrations of greenhouse gases in parts per million (ppm) and the annual emissions of those gases in billion metric tons of carbon per year (GtC/yr)—will be as widely known as the Gross Domestic Product. Well, actually more widely known, since I doubt one person in ten knows the GDP, in large part because it doesn't have much bearing on their lives. But everyone will know the global carbon dioxide level, because that single number, more than any other, will determine the fate of their children and all civilization.

Almost all of these posts are from the past two years, a time of rapid change and turmoil. My aim in putting these posts together thematically is to give readers my "straight up" take on events as they unfolded.

Overview of the Book

Averting catastrophic global warming requires completely overturning the status quo, changing every aspect of how we use energy—and doing so in under four decades. Failure to do so means humanity's self-destruction.

Unfortunately, the primary job of communicating that message to the public is in the hands of the mainstream media—the MSM, as the blogosphere refers to them. But I now prefer the term "status quo media," after reading a 2009 *Newsweek* piece that admitted the shocking, seldom-stated truth about the media elite: They have "a vested interest in keeping things pretty much the way they are." Chapter 1 shows how painfully true that is in the area of climate and clean energy.

Chapter 2 looks at the dire nature of the climate situation. If you've gotten most of your climate science filtered through the status quo media, then you are among the 99 percent of Americans who don't know what's coming if greenhouse gas emissions keep going straight up. You will hear from that small-but-growing group of scientists who are being uncharacteristically blunt with the public. I use the phrase "Hell and high water" throughout the book to describe the grim future our children and grandchildren face, since "global warming" and "climate change" are really euphemisms for what is to come if we don't reverse course soon.

Chapter 3 looks at the clean energy solutions, including "the technology that will save humanity," concentrated solar power. I debunk a widely held myth, pushed especially by those who want to delay and block climate action, that we need multiple technology breakthroughs before we can start aggressively replacing dirty energy in an affordable fashion. The truth is we already have most of the technology we need to slash emissions at very low net cost, and we will soon have the rest. If we don't start deploying existing technology now, all the new technology in the world won't be able to preserve a livable climate. The best way to lower the cost of clean energy even further is to start pushing it into the market now.

Chapter 4 looks at the other major energy problem facing the nation, the imminent peak in global production of oil. "We have to leave oil before oil leaves us," warned Dr. Fatih Birol, the chief economist at the historically staid and conservative International Energy Agency, in August 2009. I also discuss a key solution to peak oil—the plug-in hybrid electric vehicle.

Chapter 5 looks at what the Obama administration has done to jump-start the transition to a sustainable future. Ultimately, Obama's legacy will rise or fall with his ability to achieve a national and a global climate deal, since those are the two central actions needed to give humanity a serious chance of averting catastrophic climate impacts. And yet what Obama has achieved in just the first year in office is remarkable, if little remarked on by the status quo media.

Chapter 6 is a brief look at the final, painful years of President George W. Bush and Vice President Dick Cheney. If humanity doesn't preserve a livable climate, future generations will judge them far more harshly than even the most liberal critic. As a blogger, it was sometimes hard to know whether to cry or laugh at their absurdities. In this chapter, I do both.

Chapter 7 looks at the climate disinformation campaign, which is driven by conservatives—their think tanks, media, pundits, and politicians. The conservative movement as a whole has embraced policies and a demagogic message that, if successful, will be responsible for ushering in decades if not centuries of untold misery along with the intrusive government that such hardship and scarcity will entail. I also offer my explanation of "The real reason conservatives don't believe in climate science."

Chapter 8 looks deeper into those who deny climate science, who self-destructively urge humanity to delay action, and who attack all those who either articulate the science or urge immediate action. I examine the polling that makes clear "The deniers are winning, but only with the GOP." That is, the message of the deniers is really resonating only with conservatives. In this book, I reserve the term "deniers" for the professional anti-scientific disinformers. I use the term "delayers" for those who have been misled by the deniers into embracing inaction.

Chapter 9 explores how scientists, environmentalists, and progressives are not great at messaging. The scientific community in particular seems enamored by communication strategies that are counterproductive, especially in an age when the status quo media is scaling back its science reporting. You'll find some suggestions for how to improve communications, including a brief look at rhetoric, the art of persuasion.

The conclusion asks, "Is the global economy a Ponzi scheme?" This richest of all human generations has figured out how to live off the wealth of future generations. Investors (i.e., current generations) are paying themselves (i.e., you and me) by taking the nonrenewable resources and livable climate from future generations. To perpetuate the high returns that rich countries have been achieving in recent decades, we have been taking an ever greater fraction of nonrenewable energy resources (especially hydrocarbons) and natural capital (fresh water, arable land, forests, fisheries), and the most important nonrenewable natural capital of all—a livable climate. The next few years will determine whether or not we are all Bernie Madoffs.

Finally, in the afterword, I discuss what I learned from attending the big December 2009 climate conference in Copenhagen that set the stage for the big debate in this country in 2010 (and beyond) over how best to cut carbon pollution and boost clean energy.

The ultimate reason that I blog is because it's not too late. Just because the catastrophic climate changes we are headed toward will probably be irreversible for hundreds of years or longer, that doesn't mean they are unstoppable.

We are going to adopt the clean energy strategies described in this book. That is a certainty. But the question of our time is, will we do it fast enough?

Humanity has only two paths forward at this point. As President Obama said in April 2009, "The choice we face is not between saving our environment and saving our economy. The choice we face is between prosperity and decline."

Either we voluntarily switch to a low-carbon, low-oil, low-net-water use, low-net-material use economy over the next two decades or the post-Ponzi-scheme collapse will force us to do so circa 2030. The only difference between the two paths is that the first one spares our children and grandchildren and future generations untold misery and expense.

CHAPTER 1

The Status Quo Media

MEDIA COVERAGE OF GLOBAL WARMING HAS not been very good nor is it likely to improve. Historically, even the most respected newspapers have fallen into the trap of giving the same credence—and often the same amount of space—to a handful of U.S. scientists, most receiving funds from the fossil fuel industry, as they give to hundreds of the world's leading climate scientists. No surprise that much of the public has ended up with a misimpression about the remarkable strength of our scientific understanding and the need for action (see chapter 8).

The study "Balance as Bias: Global Warming and the U.S. Prestige Press" analyzed more than 600 news articles published from 1990 to 2002 in the *New York Times, Washington Post, Los Angeles Times*, and *Wall Street Journal.* The study found "significant difference between the scientific community discourse and the U.S. prestige press discourse." For instance, "53 percent of the articles gave roughly equal attention to the views that humans contribute to global warming and that climate change results exclusively from natural fluctuations."

In my blogging since mid-2006, I've found that the media coverage has not improved much. Why? One reason is that as the climate story has become a first-tier political story, more and more pieces are being written by senior political reporters, who know very little about global warming and who haven't bothered to educate themselves on what is indisputably the story of the century. Instead, they employ the horse-race perspective that dominates today's political coverage, attempting only to measure who is up and who is down. The publication on the web of the e-mails stolen from UK researchers in late 2009 allowed many media outlets to continue to mis-cover the science and give undue credence to those spreading anti-science disinformation.

Media critiques are among the most popular Climate Progress.org pieces. As the selection of posts in this chapter shows, media coverage across the board—from the science to the economics to the solutions—is still doing a grave disservice to the public.

Media Enable Denier Spin Part 2: What If the MSM Simply Can't Cover Humanity's Self-Destruction?

March 5, 2008

If those who are counseling inaction and delay succeed, billions of humans will suffer unimaginable misery and chaos, while most other species will simply go extinct.

Maybe the best one-line description of our current situation that I have read is:

> It may seem impossible to imagine that a technologically advanced society could choose, in essence, to destroy itself, but that is what we are now in the process of doing.

That's the final sentence in Elizabeth Kolbert's fine global warming book, *Field Notes from a Catastrophe*, and as I'll show in this post, it is entirely accurate.

How can the traditional media cover a story that is almost "impossible to imagine"? I don't think they can. I'll be using a bunch of quotes mostly from Andy Revkin of the *New York Times*, not because he is a bad reporter—to the contrary, he is a leading climate reporter—but because now that he has a blog, he writes far more than any other journalist on this subject and shares his thinking. A new Revkin post, "The Never-Ending Story," underscores the media's central problem with this story:

> I stayed up late examining the latest maneuver in the **never-ending tussle** between opponents of limits on greenhouse gases who are using **holes in climate science** as ammunition and those trying to raise public concern about **a human influence on climate that an enormous body of research indicates, in the worst case, could greatly disrupt human affairs and ecosystems.**

This sentence is not factually accurate (the boldface highlighting the passages at issue is mine). It would be much closer to accurate if the word "worst case" were replaced by "best case" or, as we'll see, "best case if the opponents of limits on GHGs fail and fail quickly." The worst case is beyond imagination. The word "holes" is misleading. And this isn't a "tussle"—it is much closer to being a "struggle for the future of life as we know it." And all of us—including Andy—better pray that it ain't "never-ending." Before elaborating, let me quote some more:

> One of the unavoidable realities attending global warming—a reality that makes it the *perfect* problem—is that there is **plenty of remaining uncertainty**, even as the basics have grown ever firmer (my litany: more CO_2 = warmer world = less ice = rising seas and lots of climate shifts).
>
> Some skeptics have long tried to use the uncertainty as an excuse for maintaining the status quo. **Campaigners for carbon dioxide curbs seem reluctant to acknowledge the gaps for fear that society will tune out. So the story migrates back to the edges: catastrophe, hoax. No doubt.**

This last paragraph sums up the problem for the media. As an aside, I don't know what "gaps" or "holes" Revkin is talking about, but as I will try to make clear, they don't really exist in the sense that any typical reader would expect from the context.

The "story migrates back to the edges," not because that is inherent to the story, but because that is inherent to all modern media coverage of every big issue. Let me quote *Newsweek* editor Jon Meacham from last month:

> I absolutely believe that the media is not ideologically driven, but conflict driven. If we have a bias it's not that people are socially liberal, fiscally conservative or vice versa. It is that we are engaged in the storytelling business. And if you tell the same story again and again and again—it's kind of boring.

The real climate story doesn't have much conflict: It is the growing scientific (and technological) understanding that if we don't sharply restrict greenhouse gas emissions soon, we face *catastrophe*—that is the right word, the one Kolbert sticks in her title.

The conflict is actually a political one between those who believe in government-led solutions and those who don't. This is a central point. As Revkin himself notes about the Heartland denier/disinformer conference, "The one thing all the attendees seem to share is a deep dislike for mandatory restrictions on greenhouse gases." As I explain (see chapter 7), a central reason that conservatives and libertarians reject the scientific understanding of human-caused climate change is that they simply cannot stand the solution. So they attack both the solution and the science.

It simply is not accurate to say the real edges of this debate are "catastrophe" or "hoax." Revkin and every reasonable person knows that this is no "hoax," no conspiracy of the thousands of top scientists in the world to deceive the public—that is laughable, pure disinformation from the conservatives who hate regulations. It is comparable in credibility to the claim that we never landed on the moon.

Revkin also knows, or he should know, that "catastrophe" is not the edge of the debate. Let me explain why.

The *middle* of the debate is the UN's Intergovernmental Panel on Climate Change (IPCC) reports. That is the mainstream scientific view. That is the "consensus" among our top scientists (even though that is a terrible word, since it suggests "consensus of opinion" rather than "Collective understanding of the science," which is accurate). The language is signed off on by more than 180 governments. You certainly can't call that the "edge."

This is especially true *if* we actually listen to the deniers/disinformers and don't act to reduce emissions soon. This is the central point that the media either fails to understand or refuses to communicate.

Let me make two related arguments. First, Revkin writes of "a human influence on climate that an enormous body of research indicates, in the worst case, could greatly disrupt human affairs and ecosystems." That research, summarized by the IPCC, says, for instance:

> Climate change is likely to lead to some irreversible impacts. There is medium confidence that approximately 20–30% of species assessed so far are likely to be at increased risk of extinction if increases in global average warming exceed 1.5–2.5°C (relative to 1980–1999). As global average temperature increase exceeds about 3.5°C, model projections

> suggest significant extinctions (40–70% of species assessed) around the globe.

Now wouldn't losing 40 percent to 70 percent of all species—a 3.5°C rise is a certainty if we lose the "tussle" to opponents of limits on GHGs—be a "catastrophe" by any definition of the word? And that's without even bringing in the hundreds of millions of people whose lives will be ruined by sea level rise, drought, and water shortages.

Let me go further. As I (and others, including the IPCC itself) have repeatedly explained, the "holes" or "gaps" in the IPCC work are almost exclusively omissions of hard-to-model things like carbon-cycle amplifying feedbacks and dynamic ice sheet destruction that would tend to make future impacts much worse than the IPCC models. And the actual observational record clearly shows that the climate is changing faster than the IPCC models project.

An even more important point, one that the media almost completely ignores, is that the other major source of "uncertainty" in the IPCC reports is that nobody knows for sure what quantity of greenhouse gases humans will emit this century. So the IPCC models a wide range of emissions, including some very low emissions/concentrations scenarios with relatively modest, but still severe, impacts. But the longer we listen to the do-nothing crowd, the more certain it is that we will end up with very high emissions and concentrations whose brutal impacts are all too certain.

How high?

We are currently emitting 8 billion tons of carbon a year (8 GtC/yr) and rising more than 3 percent per year—faster than the most pessimistic IPCC model. There is a little-reported bombshell buried in the footnote of the first IPCC report released last year:

> Based on current understanding of climate carbon cycle feedback, model studies suggest that to stabilise at 450 ppm carbon dioxide, could require that cumulative emissions over the 21st century be reduced from an average of approximately 670 GtC to approximately 490 GtC. Similarly, to stabilise at 1000 ppm this feedback could require that cumulative emissions be reduced from a model average of approximately 1415 GtC to approximately 1100 GtC.

"Oh my dear God" is the appropriate response to the final sentence, if you oppose greenhouse gas regulations—or if you worry that those who do will maintain enough credibility/influence in the media and in Washington, DC, that they (continue to) succeed in stalling action—or if you actually are a member of the media who treat those opponents as if they had a scientifically or morally defensible position.

On our current emissions pace, we will be at 11 billion metric tons of carbon a year (GtC/yr) around 2020 and still rising! That means, if the do-nothing side wins—or even if they just partially win (by limiting government controls to ones that lead to average emissions of 11 GtC/yr for the century)—then the planet's carbon dioxide concentrations, feedbacks included, are headed to 1000 ppm!

Let me repeat, if the do-nothing or do-little side wins, we face 1000 ppm atmospheric concentration of CO_2—a quadrupling from preindustrial levels—if not higher. That is *not* the worst case; that isn't even business as usual if the disinformers win: Stabilizing at 1000 ppm still requires a lot of government-led effort that conservatives almost universally disdain.

Scientists rarely even bother modeling the impacts of 1000 ppm because "catastrophe" doesn't begin to describe the impacts. We are talking average global temperatures some 5.5°C higher—yes 9°F higher (and much higher than that on northern land masses like the continental United States)—in any case, far higher than the last time the planet had no ice whatsoever and sea levels were more than 250 feet higher. The ocean would be rendered virtually lifeless. Dust Bowls would engulf one third of the habited land mass of the planet. This is not "global warming" or "climate change," it is Hell and high water (see chapter 2). Few scientists have, perhaps until very recently, seriously considered that humanity would be so mindlessly self-destructive that 1000 ppm would be a possible outcome.

To repeat the bottom line: If those who council inaction and delay win, then there is no uncertainty about our future, no gaps, no holes, nothing but decades if not centuries of misery for billions and billions of humans and outright extinction for most other species.

I get that the media treats so-called alarmists with skepticism. I sort of understand why Revkin writes this weekend about the conference of skeptics "trying hard to prove that they had unraveled the established science showing that humans are warming the world in potentially disruptive ways," as opposed to more accurate statements like "in potentially catastrophic ways" or "in ways that will be catastrophic if we actually listen to the skeptics."

I understand that much of the traditional media either hasn't taken the trouble to figure out what's going to happen to humanity if the anti-government crowd wins—or, for those in the media who know, they feel they just can't keep beating the public over the head with the painful truth.* But every time they do a story with a different, blander spin, they undermine the urgent need for action. Every time they say there is a middle ground, they push us closer to the certain catastrophe of inaction. I think that qualifies as tragic irony.

So yes, it appears to me that today's media simply can't cover humanity's self-destruction. When historians write about this time—very, very bitterly, no doubt, if we have forced them to suffer through Hell and high water—the media will get assigned plenty of blame for sins of omission, though obviously not as much blame as those who were actively working to spread disinformation and block action.

I will end with a quote about the journalism of my father's time—Dad was an investigative journalist in addition to being a newspaper editor for decades—from Edward R. Murrow, *See It Now*, March 9, 1954:

> This is no time for men who oppose Senator McCarthy's methods to keep silent. We can deny our heritage and our history, but we cannot escape responsibility for the result.

---

* I guess that's what blogs are for.

Must-Read Newsweek Stunner: Why the "Status Quo" Establishment Media's Coverage of Global Warming Is So Fatally Useless, Part 1

March 30, 2009

Averting catastrophic global warming requires completely overturning the status quo, changing every aspect of how we use energy—and doing so in under four decades. Failure to do so means humanity's self-destruction. Media coverage of the problem and the solution has been dreadful. But why? In his new cover story on Paul Krugman, *Newsweek*'s Evan Thomas unintentionally provides the answer—the shocking, unstated truth about the media elite: They have "a vested interest in keeping things pretty much the way they are."

Assuming we don't spend the small fraction of GDP per year needed to avert catastrophe (see chapter 3), future generations who are puzzled about our fatal myopia need look no further for explanation than Thomas's full remarks. He begins with the amazing admission, "If you are of the establishment persuasion (and I am)," and continues with words that should be emblazoned across journalism schools around the country and read out loud at every Ivy league college graduation:

> By definition, establishments believe in propping up the existing order. Members of the ruling class have a vested interest in keeping things pretty much the way they are. Safeguarding the status quo, protecting traditional institutions, can be healthy and useful, stabilizing and reassuring. But sometimes, beneath the pleasant murmur and tinkle of cocktails, the old guard cannot hear the sound of ice cracking. The in crowd of any age can be deceived by self-confidence. . . .

Thomas was writing about the current economic crisis, but his words apply far better to the global Ponzi scheme (see the conclusion to this book). Indeed, his words could not more ironically apply to the catastrophic global warming that he and his establishment buddies are all but blind to:

> . . . the old guard cannot hear the sound of ice cracking.

This might just be an epitaph for modern human civilization, given reports in 2008 that 2 trillion tons of land ice were lost since 2003, and the rate of Greenland summer ice loss tripled the 2007 record. Startling new sea-level rise research indicates a rise of 3 to 7 feet "most likely" by 2100.

Glenn Greenwald's column in *Salon*, "*Newsweek*'s unintentionally revealed, central truth," put me on to this story. He notes that it is not just Thomas, but "also most of his media-star colleagues," who are "of the establishment persuasion." He concludes:

> One day in the near future, Thomas should have a luncheon or perhaps a nice Sunday brunch at his home, invite over all of his journalist friends who work in the media divisions of our largest corporations, and they should spend fifteen minutes or so assembling these sentences together, and then examine what these facts mean for the actual role played by establishment journalists, the functions they fulfill, whose interests they serve, and the vast, vast disparities between (a) those answers and (b) the pretenses about their profession and themselves which they continue, ludicrously, to maintain. . . .

Also, in the name of consumer protection, television news shows and the largest newspapers ought to place that paragraph by Thomas as a warning at the top of every product they produce.

Hear! Hear!

David Broder Is the Sultan of the Status Quo, Stenographer of Those Centrists Who Are Fatally Uninformed about Global Warming, Part 2

April 14, 2009

> The hottest places in hell are reserved for those who, in times of great moral crisis, maintain their neutrality.

That is attributed to Dante, but applies best to the Washington establishment, especially one David Broder.

Part 1 looked at why the establishment media's coverage of global warming is so fatally useless. *Newsweek*'s Evan Thomas unintentionally provided the answer—the shocking, unstated truth about the media elite: They have "a vested interest in keeping things pretty much the way they are."

But Evan Thomas is a B-list establishment journalist compared to the dean of the DC press corps—David Broder. In two recent columns, Broder has combined a scientifically uninformed position on climate with remarkably flawed political analysis designed to support his position. Let's start with the absurdities in his most recent piece: "Why the Center Still Holds":

> Once political independents, who like the idea of clean air, grasped that cap-and-trade would mean a big tax increase for them, Republican opposition was reinforced and Democratic support weakened to the point that the Obama plan may already be doomed this year.

Huh? Cap-and-trade doesn't mean a big tax increase. That would be a right wing talking point that they beat to death again and again to sucker . . . well, it's obvious who they are trying to sucker. So much for Broder being a "centrist" or an independent.

I guess it bears repeating over and over again that the combination of aggressive investment in energy efficiency and the president's plan to return most of the auction revenues to the public means the majority of the public is held harmless—and indeed can actually lower their combined energy and tax bill if they adopt energy efficiency with the help of their utility or the federal government's low-income weatherization program (see chapter 3).

By focusing only on the cost of action, and ignoring entirely the cost of inaction, Broder is yet another poster child for the searing critique award-winning journalist Eric Pooley did for Harvard (discussed later in this chapter).

Second, who the heck didn't "grasp" a long time ago that cap-and-trade would raise the price of (dirty) energy? And when did many Republicans ever support action—you can go back several years to earlier climate bills and find very little support. Republican opposition couldn't possibly be "reinforced" given how many have been dead set against any action whatsoever for years.

I grant that Democrats have done a lousy job explaining that a cap-and-trade never belonged in the budget in the first place. Again, climate legislation was never going to be easy, but in any case nothing that has happened recently suggests Democratic support is any weaker—or stronger—than it was two years ago.

The really sad thing about Broder is that in two columns on the subject, he never bothers even mentioning a single reason why action on energy and climate is needed. In the "End of the Honeymoon," he writes:

> I think the shift began when Obama moved beyond the stimulus bill to his speech to the joint session of Congress and his budget message. For the first time, the full extent of his ambitions for 2009 became clear—not just stopping and reversing the steep slide in the economy but also launching highly controversial efforts in health care, energy, and education.
>
> Each of those issues has a history in Washington—a history marked by congressional gridlock and legislative frustration.

In Broder's world of uninformed centrism, if an issue has a history of gridlock and legislative frustration, then it is "highly controversial" and any president who tries to address these absolutely crucial issues is reaching too far.

Again, note that he never bothers to engage the substance of the issue. The media establishment doesn't care about substance. It cares only about the status quo. To repeat what Thomas wrote:

> By definition, establishments believe in propping up the existing order. Members of the ruling class have a vested interest in keeping things pretty much the way they are. Safeguarding the status quo, protecting traditional institutions, can be healthy and useful, stabilizing and reassuring. But sometimes, beneath the pleasant murmur and tinkle of cocktails, the old guard cannot hear the sound of ice cracking. The in crowd of any age can be deceived by self-confidence. . . .

That would be David Broder.

Two final points. Broder drags out seriously flawed political analysis to attack Obama as polarizing:

> As for the voters, the Pew Research Center reported this month on a survey that showed the partisan gap in Obama's job approval scores is the widest in contemporary history. He rated a thumbs-up from 88 percent of the Democrats and only 27 percent of the Republicans in the poll—a gap of 61 points.
>
> At a comparable point in their first terms, the gaps for George W. Bush and Bill Clinton were only 51 and 45 points, respectively.

Uhh, even the *Washington Post*'s own political reporter, Dan Balz, explained a key reason for that change in statistics:

> Another factor is that, in a shrinking Republican Party, conservatives hold more sway—and they are most likely to disapprove of a Democratic president's performance. Exit polls show that 64 percent of Republicans who voted in November called themselves conservatives. That compares with 54 percent in 2000 and 49 percent in 1992.

So the main reason Obama *appears* to be more polarizing in terms of a ten-point bigger gap in relative favorability among Democrats versus Republicans (compared to Bush) is that Republicans have gotten ten-points more conservative.

But that analysis would get in the way of Broder's attack on Obama as someone who is polarizing because Obama attacks the status quo because he wants to avoid catastrophic global warming and deal with our unsustainable use of oil.

Or how about this from Broder's first piece:

> Congress has taken note of the way Obama backed down from his anti-earmark stance, a clear signal that he is leery of any showdown with the lawmakers. Despite his popularity, Obama is not an intimidating figure, and so he can expect to be tested time and again.

So let me see. First, Broder attacks Obama for overreaching by trying to address "highly controversial" issues like energy, even though that is precisely what a president should use his popularity for. Then Broder attacks Obama for not using his popularity for a "showdown" with lawmakers on the trivial earmark issue, which comprises about 2 percent of the budget.

Further note to Broder: Even if Obama cut out all of the earmarks, it wouldn't save a penny of taxpayer money since the earmarks just cordon off parts of the budget—Congress would still use the money for other spending.

But the bottom line is clear. Broder thinks Obama should have burned up his popularity on a trivial process issue (earmarks), but should stay far away from the nation's substantive problems like health care or energy, since that is only what polarizing politicians pursue.

The status quo approach of the David Broders of the Washington establishment are the road to Hell and high water, also known as catastrophic global warming.

CNN, ABC, WashPost, AP, Blow Australian Wildfire, Drought, Heatwave "Hell (and High Water) on Earth" Story—Never Mention Climate Change

February 10, 2009

If the U.S. media refuse to make the connection between record-breaking wildfire, drought, and heat waves and human-caused global warming, why would anyone be surprised if the U.S. public doesn't put it as a higher priority or make the connection itself?

Australia knows it's facing climate-driven impacts that threaten it with complete collapse. AFP (French international media) gets this: "Australian wildfire ferocity linked to climate change: experts." So does Reuters' climate change correspondent in Asia: "Australia fires a climate wake-up call: experts."

I saw the CNN and ABC stories, and you can read the AP's stories, which have been published in the *Washington Post* and *New York Times* (though the *NYT* redeemed itself, as we will see). The media love a good calamity of Biblical proportion: "Hell in all its fury has visited the good people of Victoria," Prime Minister Kevin Rudd told reporters as he toured the fire zone on Sunday.

But for the vast majority of the U.S. media, you won't find any mention of a global-warming-driven heat wave or drought as the underlying cause of this calamity. ABC's Charles Gibson said "the worst wildfires in Australian history" were "part natural disaster" and "part manmade crime" because arson is suspected in some of the fires. No, Charlie, the natural disaster is not entirely natural, so this is mostly a manmade crime.

The AP story in the *Washington Post* ends lamely:

> Wildfires are common during the Australian summer. Government research shows about half of the roughly 60,000 fires each year are

> deliberately lit or suspicious. Lightning and people using machinery near dry brush are other causes.

Contrast that to AFP, which leads their story with:

> Australia is naturally the most fire-prone continent on Earth but climate change appears to be making the wildfires that regularly sweep across the country more ferocious, scientists said Monday.

The AFP story also contains the warning:

> Research by the Bureau of Meteorology and the government science organisation CSIRO predicts the number of days when bushfires pose an extreme risk in southeastern Australia could almost double by 2050 under a worst-case climate change scenario.

The Reuters Asia story notes:

> Some analysts say the fires were predictable and that climate scientists have been warning for years about Australia's vulnerability to rising temperatures and declining rainfall across much of the nation's south.
>
> "I would compare this current bushfire event to one of the ghosts in Dickens's *Christmas Carol* that visits Scrooge and showed him what his future would be like if he didn't change his ways," said professor Barry Brook, director of the Research Institute for Climate Change and Sustainability at the University of Adelaide.

Hmm. Maybe Prof. Brook reads ClimateProgress.org (I had made the same exact analogy).

As an aside, while one part of Australia is burning like "Hell in all its fury," some readers note the incredible flooding in Queensland: "60 percent of the state is under water!" As one story noted, "Far north Queensland in chaos after heavy flooding."

So, yes, the point of this semi-digression is that Australia is simultaneously experiencing Hell and high water.

The *New York Times* ran a much better story today than their earlier AP-inspired stories, no doubt because "Andrew C. Revkin contributed reporting from New York." The story noted:

> The firestorms and heat in the south revived discussions in Australia of whether human-caused global warming was contributing to the

> continent's climate woes of late—including recent prolonged drought in some places and severe flooding last week in Queensland, in the northeast.
>
> Climate scientists say that no single rare event like the deadly heat wave or fires can be attributed to global warming, but the chances of experiencing such conditions are rising along with the temperature. In 2007, Australia's national science agency published a 147-page report on projected climate changes, concluding, among other things, that "high-fire-danger weather is likely to increase in the southeast."
>
> The flooding in the northeast and the combustible conditions in the south were consistent with what is forecast as a result of recent shifts in climate patterns linked to rising concentrations of greenhouse gases, said Kevin Trenberth, a scientist at the United States National Center for Atmospheric Research.

So kudos to the *NYT* for a much more complete story then most everyone else in the US media. Too bad the story is on page A9, the paragraphs cited are at the very end, and the headline is "Australia Police Confirm Arson Role in Wildfires."

Must-Read Study: How the Press Bungles Its Coverage of Climate Economics—"The Media's Decision to Play the Stenographer Role Helped Opponents of Climate Action Stifle Progress."

January 25, 2009

One of the country's leading journalists has written a searing critique of the media's coverage of global warming, especially climate economics.

How Much Would You Pay to Save the Planet? The American Press and the Economics of Climate Change is by Eric Pooley for Harvard's prestigious Joan Shorenstein Center on the Press, Politics, and Public Policy. Pooley has been managing editor of *Fortune*, national editor of *Time*, *Time*'s chief political correspondent, and *Time*'s White House correspondent, where he won the Gerald Ford Prize for Excellence in Reporting. Before that, he was senior editor of *New York* magazine.

In short, Pooley has earned the right to be heard. Journalists and senior editors need to pay heed to Pooley's three tough conclusions about how "damaging" the recent media of the climate debate has been:

> The press misrepresented the economic debate over cap and trade. It failed to recognize the emerging consensus . . . that cap-and-trade would have a marginal effect on economic growth and gave doomsday forecasts coequal status with nonpartisan ones. . . . The press allowed opponents of climate action to replicate the false debate over climate science in the realm of climate economics.
>
> The press failed to perform the basic service of making climate policy and its economic impact understandable to the reader and allowed opponents of climate action to set the terms of the cost debate. The argument centered on the short-term costs of taking action—that is, higher electricity and gasoline prices—and sometimes assumed that doing nothing about climate change carried no cost.
>
> Editors failed to devote sufficient resources to the climate story. In

> general, global warming is still being shoved into the "environment" pigeonhole, along with the spotted owls and delta smelt, when it is clearly to society's detriment to think about the subject that way. It is time for editors to treat climate policy as a permanent, important beat: tracking a mobilization for the moral equivalent of war.

Precisely.

Pooley is one of the few major American journalists who understands that global warming is the story of the century—and if we don't reverse our emissions path soon, it will likely be the story of the millennium, with irreversible impacts lasting for many, many centuries.

In a conversation Saturday, Pooley told me, "I think this is the only story going forward." That's why, although he remains a contributor to *Time* magazine, he is devoting most of his time now to researching and writing a book on the politics and economics of climate change.

The first step for Pooley was an analysis of media coverage during the last fifteen months. In a long introduction to the different roles reporters can play, Pooley notes that "being a referee is harder than being a stenographer because it requires grappling with the substance of an issue in a way that many time-pressed journalists aren't willing or able to do."

He decided to examine media coverage surrounding the 2008 Senate debate over the climate bill put forward by John Warner (R-VA) and Joseph Lieberman (I-CT):

> News coverage of the Lieberman-Warner debate included some shoddy, one-sided reporting and some strong work that took the time both to dive into the policy weeds—evaluating the economic assumptions used by the various players—and step back to portray those players as combatants in a war for public opinion. But most of the reporting was bad in the painstakingly balanced way of so much daily journalism—two sides, no real meat.

He then explains his research:

> My analysis of news articles published in national and regional newspapers, wire services, and newsmagazines between December 2007 and June 2008 suggests that for most reporters covering this story, the

> default role was that of stenographer—presenting a nominally balanced view of the debate without questioning the validity of the arguments, sometimes even ignoring evidence that one side was twisting truth. Database searches yielded a sample of forty published news and analysis stories that explored the cost debate in some detail (see appendix). Of these, seven stories were one-sided. Twenty-four stories were works of journalistic stenography. And nine stories attempted, with varying degrees of success, to move past the binary debate, weigh the arguments, and reach conclusions about this thorny issue.

The bottom line:

> The media's collective decision to play the stenographer role actually helped opponents of climate action stifle progress.

He makes another interesting point, one I would not have expected from a journalist:

> Mainstream news organizations have accepted the conclusions of the IPCC but have not yet applied those conclusions to the economic debate. The terms of that debate have been defined by opponents of climate action who argue that reducing emissions would "cost too much." So the battle has been fought over the short-term price of climate action and its impact on GDP, while overlooking an extremely important variable, the long-term costs of inaction and business as usual.

Although Pooley doesn't make the point, the problem he identifies is compounded by the fact that the mainstream economic community also overestimates the cost of action and underestimates the cost of inaction, a central point of my ongoing series on voodoo economists.

That means when the media goes out looking for a well-known climate economist to quote in an article, they typically end up with someone who doesn't understand the scientific urgency and those who misunderstand the economics.

If you really want to understand the fact that even a very strong cap-and-trade bill "would have a marginal effect on economic growth," the best places to go are the International Energy Agency, the IPCC, and McKinsey & Company (see chapter 3).

Pooley's whole paper is a must-read, especially for advocates of climate action. Yes, the media bears much culpability for the fact that, as Pooley says, "the tipping point for climate action has not yet been reached." But so do scientists, environmentalists, and progressives. The general state of our messaging remains lousy (see chapter 9).

One clear message from this study is that the climate science activists need to do a better job of spelling out the cost of inaction. Until that cost is clear to the public, media, and policy makers, the country will never be able to mobilize to do what is needed to preserve a livable climate.

Memo to *Wall Street Journal*: You Can Do Better than "Greenhouse Gases, Which Are Believed to Contribute to Climate Change"

April 2, 2009

The media misinforms the public about climate science in many different ways. One is by publishing long-debunked disinformation over and over again (see George Will and Charles Krauthammer pieces in chapter 7).

But misinformation can be as damaging as disinformation. Consider this March 27 *Wall Street Journal* piece, "Climate Talks Look to U.S. Role," by Leila Abboud and Stephen Power (e-mailed me by a sharp-eyed reader). It contains this pointlessly hedged sentence:

> The U.S., under the Bush administration, didn't ratify the Kyoto treaty, and China and other developing countries such as India and Brazil aren't obligated under the treaty to restrict emissions of greenhouse gases, which are believed to contribute to climate change.

I think we are at least one decade, if not two decades, past a time when the words "are believed to" are justified.

Note to Abboud and Power: Why exactly do you think they are called greenhouse gases?

This hedge is especially pointless and misinforming because of the second hedge—"contribute to."

Back in 2001, the Intergovernmental Panel on Climate Change (IPCC)—the world's top climate scientists who periodically review the scientific literature and publish reports that every major government signs off on word for word—wrote:

> An increasing body of observations gives a collective picture of a warming world and other changes in the climate system. . . . There is new and stronger evidence that most of the warming observed over the last fifty years is attributable to human activities.

Based on increasingly strong scientific evidence, the IPCC strengthened its conclusion in 2007, as the *New York Times* explained:

> The world's leading climate scientists said global warming has begun, is *very likely* caused by man, and will be unstoppable for centuries. . . . The phrase *very likely* translates to a more than 90 percent certainty that global warming is caused by man's burning of fossil fuels. That was the strongest conclusion to date, making it nearly impossible to say natural forces are to blame.

Perhaps the *Wall Street Journal* might catch up with the scientific understanding and write some variation of:

> . . . emissions of greenhouse gases, which cause climate change.

NYT's Matt Wald Blows the "Alternative and Renewable Energy" Story, Quotes Only Industry Sources, Ignores Efficiency and Huge Cost of Inaction

March 29, 2009

I have known the *New York Times* energy reporter, Matt Wald, for fifteen years, and generally think he is pretty good. But he has published perhaps the most flawed, inaccurate, and indefensible article of his career.

Wald's piece could also be a poster child for award-winning journalist Eric Pooley's searing critique of the media's coverage of climate economics earlier in this chaper.

And, amazingly, as we will see, a report by one of Wald's two industry sources completely disagrees with the report by the other industry group Wald cites! In fact, new concentrated solar thermal power (see in chapter 3) is already competitive with new gas-fired generation and could well have better economics by 2015.

The first flaw is that Wald completely ignores the lowest-cost electricity strategy—energy efficiency—even though the article's headline is "Cost Works against Alternative and Renewable Energy Sources in Time of Recession," and a major point of the piece is that "curbing carbon dioxide emissions—a central part of tackling climate change—will almost certainly raise electricity prices."

Wald never tells the reader that until the economic collapse, traditional sources of power had been rising much faster in cost than alternatives. He also never mentions that efficiency, which costs two cents to four cents a kilowatt hour (not counting ancillary benefits, including no need for new transmission), is the only new source of power that is both pollution-free and far cheaper than current electricity rates (see chapter 3).

The media simply needs to start talking more about electricity *bills*, which encompass efficient use of energy, than about electricity *rates.*

Second, just as Pooley specifically warns against, Wald cites only industry sources for cost—and, surprise, surprise!—they have absurd and indefensible numbers. Indeed, the clearest evidence article of bias is the utterly insupportable cost estimate for nuclear power Wald cites from a Black and Veatch study, "a new nuclear reactor, 10.8 cents" (per kilowatt-hour).

Matt, say it ain't so. That number is beyond unsupportable. There is not a utility or nuclear power plant provider in the country who would guarantee 10.8 cents/kwh in a Public Utility Commission (PUC) hearing. You would have trouble finding one that would guarantee *twice that rate* in year one of operation.

Let's remember that "Turkey's only bidder for a first nuclear plant offers a price of 21 cents per kilowatt-hour." Moody's—a far less biased source than Wald cites—puts new nuclear at over 15 c/kwh. Earlier this year, *Time* wrote "new nuclear energy is on track to cost 15¢ to 20¢ per kilowatt-hour," and I published a detailed cost study that put it at 25 c/kwh or more.

After Progress Energy tripled its cost estimate for its new twin 1,100 MW nukes to $17 billion last year, it warned state regulators its estimate for its planned nuclear facility is "nonbinding" and "subject to change over time" (see my May 2008 report, "The Self-Limiting Future of Nuclear Power").

I recently debated Anndria Gaerity, director for nuclear development at PSEG Power, and as anybody who was there can attest, she did not dispute my cost numbers nor the fact that no utility in the country would guarantee lower costs in a PUC rate hearing.

You can safely ignore all of Matt Wald's numbers in the piece.

Third, Wald ignores numerous other credible sources that give very different cost estimates—including a study by one of Wald's own sources! He cites Black and Veach that "A modern coal plant of conventional design, without technology to capture carbon dioxide before it reaches the air, produces at about 7.8 cents a kilowatt-hour."

Well, I doubt you could find a coal utility to guarantee that price at a

PUC hearing. Moody's says coal would cost more than 11 cents/kwh—and that assumes no cost for emitting CO_2, another flaw in the article I will come back to.

Worse, Wald cites another industry study for costs in 2015:

> The Electric Power Research Institute, a nonprofit consortium financed by investor- and publicly owned utilities, predicted in November that even for plants coming on line in 2015, wind energy would cost nearly one-third more than coal and about 14 percent more than natural gas. The cost of solar thermal electricity, made by using the sun's heat to boil water and spin a turbine, would be nearly three times that of coal and more than twice that of natural gas. (It would be almost double the cost of wind energy, too.)

Not.

According to a 2008 Sandia National Laboratory presentation, concentrated solar thermal electric power (CSP) costs are projected to drop to 8 to 10 cents per kilowatt hour when capacity exceeds 3,000 MW. The world will probably have double that capacity by 2013. A 2006 report by the Western Governors Association makes the same point, "that, with a deployment of 4 GW, total nominal cost of CSP electricity would fall below 10¢/kWh." And that deployment will likely occur before 2015. Indeed, the report noted the industry could "produce over 13 GW by 2015 if the market could absorb that much."

Or consider work done for the California Public Utility Commission (CPUC) on how to comply with the AB32 law (California's Global Warming Solutions Act). They put California solar thermal at 12.7 to 13.6 cents/kWh (including six hours of storage capacity).

And one comprehensive collection of different CSP cost estimates notes that in *Energy Technology Perspectives* (2008), "the International Energy Agency says that CSP plants *under construction* are expected to generate electricity at between 12.5 and 22.5 US cents/kWh."

Finally, if you really want to see how one-sided and unsupportable Wald's analysis is, read the 2006 report "Economic, Energy, and Environmental Benefits of Concentrating Solar Power in California," for the National Renewable Energy Laboratory," by . . . wait for it . . . Black and Veatch:

> A comparison of the levelized cost of energy (LCOE) revealed that the LCOE of $148 per MWh [14.8 cents per kwh] for the first CSP plants installed in 2009 is competitive with the simple cycle combustion turbine at an LCOE of $168 per MWh, assuming that the temporary 30 percent Investment Tax Credit is extended.

The tax credit was extended eight years in the bailout bill. But it gets better, since this analysis was really aimed at 2015 costs:

> CSP plants installed in 2015 are projected to exhibit a delivered LCOE of $115/MWh, compared with $168/MWh for the simple cycle combustion turbine and $104/MWh for combined cycle plants. At a natural gas price of about $8 per MMBtu, the LCOE of CSP and the combined cycle plants at 40 percent capacity factor are equal.

And that is without a carbon price.

So I'm afraid Wald needs much better sources.†

Fourth, we are long past the time that a serious reporter can write an article about the cost of carbon dioxide emissions and never bother once to mention any of the costs of not doing so.

Seriously, traditional media, if your cost-of-combating-climate-change articles repeat industry cost estimates, ignore obvious cost mitigation strategies that those writing climate bills have embraced, and never mention the cost of inaction—you are acting like industry flacks.

The end of Wald's piece is unintentionally ironic:

> No one likes higher bills. But the pain might not be shared equally: despite modest rate breaks for low-income customers, poor people spend a higher portion of their income on electricity than the rich.

First off, Wald inappropriately switches from talking about prices to bills here. Because of the aggressive energy efficiency strategies that Obama and Congress are pursuing, bills can stay relatively flat even as rates go up. But Wald has omitted any discussion of efficiency, so he should not have used the word "bills" here.

Second, regressivity exists only if the country doesn't take the actions

† I'm not saying that apples-to-apples electricity cost comparisons are easy, especially projections six years from now. Only that Wald picked those that are one-sided and indefensible.

Obama and Congress have promised. But they have significantly ramped up funding for low-income weatherization, and they plan to take revenues from a cap-and-trade and return them to middle- and lower-income people.

> "There are great benefits to the use of alternative energy," said Jonathan Mir, co-head of the North American power utilities group at the investment bank Lazard.
>
> But if Congress neglects the social issue, Mr. Mir said, a change in policy could fall hardest on those without a safety net.

Again, why is Wald quoting some industry source when Congress has already begun acting on the social issue? If Wald is going to write these incredibly stovepiped articles, where he doesn't talk about global warming impacts or do any political reporting, then his articles will be less than helpful, and filled with biased quotes. But I love the final line.

"If it is deployed in an uneconomic way," he said, "it is quite regressive in nature."

In nature, what is regressive is a species using up all of the planet's nonrenewable resources (water, fisheries, arable land, and so on), destroying a livable climate, and generally accelerating the extinction of most species on the planet. That is about as uneconomic as you get.

The *Washington Post* Goes Tabloid, Publishes Second Falsehood-Filled Op-Ed by Sarah Palin in Five Months—On Climate Science and the Hacked E-Mails!

Palin Jumps from Birther to Flat Earther

December 8, 2009

It is no longer possible to hide the decline of a once-great newspaper, no longer possible to hide the decline of the paper that broke the Watergate story, but is now hanging itself on the Climategate story.

The newspaper that just editorialized "Many—including us—find global warming deniers' claims irresponsible" has published a grotesquely irresponsible and falsehood-filled piece on climate science and the hacked e-mails by that leading light of science, ex-governor Sarah Palin. This is a woman that recently embraced the fact-free birthers who question Obama's U.S. birth— despite all the evidence that renders their position untenable.

Palin is so practiced at repeating falsehoods—even in her supposed area of expertise (energy)—that during last year's presidential campaign, the *Washington Post* itself gave her its highest (which is to say lowest) rating of "Four Pinocchios" for continuing to "to peddle bogus [energy] statistics three days after the original error was pointed out by independent fact-checkers." And then in July, the *Post* let her publish a falsehood-filled piece attacking climate action and clean energy.

And now they publish this unmitigated tabloid nonsense:

> The e-mails reveal that leading climate "experts" deliberately destroyed records, manipulated data to "hide the decline" in global temperatures. . . .

No, the emails don't reveal that.

Seriously, does the *Post* have any evidence that records were deliberately destroyed? In fact, that is a right-wing myth debunked weeks ago (see my blog post "Santer, Jones, and Schneider Respond to CEI's Phony Attack on the Temperature Record").

Is the *Post* the least bit concerned that the "hide the decline" e-mail was not about any nefariously "manipulated data"—everything was done in plain sight—and it was not about hiding a decline in global temperatures, but involved one small dataset. As Prof. Phil Jones of the Climatic Research Unit (CRU) at the UK's University of East Anglia himself explained at length:

> The use of the term "hiding the decline" was in an e-mail written in haste. CRU has not sought to hide the decline. Indeed, CRU has published a number of articles that both illustrate, and discuss the implications of, this recent tree-ring decline, including the article that is listed in the legend of the WMO Statement figure. It is because of this trend in these tree-ring data that we know does not represent temperature change that I only show this series up to 1960 in the WMO Statement.

As the Union of Concerned Scientists explained, the phrase "refers to omitting data from some Siberian trees after 1960. This omission was openly discussed in the latest climate science update in 2007 from the IPCC, so it is not 'hidden' at all."

On what basis does the *Post* allow Palin to assert:

> That's not to say I deny the reality of some changes in climate—far from it. I saw the impact of changing weather patterns firsthand while serving as governor of our only Arctic state. I was one of the first governors to create a subcabinet to deal specifically with the issue and to recommend commonsense policies to respond to the coastal erosion, thawing permafrost, and retreating sea ice that affect Alaska's communities and infrastructure.
>
> But while we recognize the occurrence of these natural, cyclical environmental trends, we can't say with assurance that man's activities cause weather changes.

In fact, we can say with assurance that man's activities cause the very climate change she points to—thawing permafrost and retreating sea ice. That is precisely what the IPCC's 2007 review of the scientific literature concluded, that there is a better than 90 percent chance humans are the cause of most of the recent warming:

> Most of the observed increase in global average temperatures since the mid-twentieth century is very likely due to the observed increase in anthropogenic greenhouse gas concentrations.

Is Palin a scientist? Does the *Post* simply allow anybody to make anti-scientific assertions? Here are some previous ClimateProgress.org blog posts:

- *Washington Post* Reporters Take Unprecedented Step of Contradicting Columnist George Will in a News Article
- In a Blunder Reminiscent of Janet Cooke Scandal, the *Washington Post* Lets George Will Reassert All His Climate Falsehoods Plus Some New Ones
- The *Washington Post*, Abandoning Any Journalistic Standards, Lets George Will Publish a Third Time Global Warming Lies Debunked on Its Own Pages
- Will the *Washington Post* Ever Fact-Check a George Will Column?
- Exclusive Interview with Dr. Mojib Latif, the Man Who Moved George Will to Extreme Disinformation
- Memo to *Post*: If George Will Quotes a Lie, It's Still a Lie

Never mind.

It is ironic that Palin set up a subcabinet to study responses to retreating sea ice, but proudly states in the op-ed:

> As governor of Alaska, I took a stand against politicized science when I sued the federal government over its decision to list the polar bear as an endangered species despite the fact that the polar bear population had more than doubled. I got clobbered for my actions by radical environmentalists nationwide, but I stood by my view that adding a healthy species to the endangered list under the guise of "climate change impacts" was an abuse of the Endangered Species Act.

Both she and the *Post* seem unaware of the fact that the polar bear has little chance of surviving once its primary habitat is melted away (see my blog posts, Will Polar Bears Go Extinct by 2030? Part I and Part II).

And, then the *Post* lets Palin assert without citing a single source on her behalf, "We can say, however, that any potential benefits of proposed emissions reduction policies are far outweighed by their economic costs." Again, that statement has no basis in fact—as one major analysis found, "Waxman-Markey clean air, clean water, clean energy jobs bill creates $1.5 trillion in benefits."

For the record, here's what serious media outlets and journals think of the e-mail story:

- *Nature* Editorial: "Nothing in the E-Mails Undermines the Scientific Case That Global Warming Is Real—Or That Human Activities Are Almost Certainly the Cause."
- *Washington Times*: "Stolen E-Mails Mean Less Than They Seem"
- Reuters: "ANALYSIS-Hacked Climate E-Mails Awkward, not Game Changer"
- Hacked Climate E-Mail Rebutted by Scientists
- *Time*: "The Truth Is That the E-Mails, While Unseemly, Do Little to Change the Overwhelming Scientific Consensus on the Reality of Man-Made Climate Change."

In a desperate effort to save itself in a dying industry, the *Post* has morphed into a tabloid newspaper.

CHAPTER 2

Uncharacteristically Blunt Scientists

IN NOVEMBER 2007, THE WORLD'S TOP climate scientists released their final report synthesizing all their work on climate science, impacts, and solutions. The situation was so dire that the leader of the UN effort, Rajendra Pachauri, said, "If there's no action before 2012, that's too late. What we do in the next two to three years will determine our future."

Far from being an alarmist, Pachauri was specifically chosen as Intergovernmental Panel on Climate Change chair in 2002 after the Bush administration waged a successful campaign to have him replace the outspoken Dr. Robert Watson, who was opposed by fossil fuel companies like ExxonMobil. It is knowledge of the facts, not one's personal politics, that make people climate alarmists or, as I call them, climate science realists.

According to Technorati.com, as of one month later, the synthesis report had some 265 blog reactions, whereas the August 24 YouTube video of Miss Teen South Carolina struggling to explain why a fifth of Americans can't locate the United States on a world map had more than 5,300 blog reactions. Hmmm. Perhaps these two things are related.

The poor coverage is partly due to the IPCC's own media naiveté. It doesn't put a lot of thought into publicizing its reports; heck, it released its final synthesis on November 17—a Saturday!—in Valencia, Spain. Not exactly the best way to get attention from the most intransigent and important audience: Americans. And the IPCC refuses to tell the world simply and plainly what will happen if we keep taking little or no action.

With such poor notice for a seminal document, more than 200 scientists then took the remarkable step of issuing a plea at the United Nations climate change conference in Bali in December 2007. Global greenhouse gas emissions, they declared, "must peak and decline in the next ten to fifteen years, so there is no time to lose." The Associated Press headline on the statement was "Scientists Beg for Climate Action." What will we drive climate scientists to next? A hunger strike?

Ironically, the 2007 summary report *underestimated* likely climate impacts for two reasons. First, its climate models ignore most of the positive or amplifying feedbacks in the climate system that tend to accelerate warming. Second, the report included a variety of emissions scenarios that assume very, very aggressive national and global action to quickly and sharply reduce greenhouse gas emissions. Much of the research discussed in this chapter takes a more realistic view of the climate system.

Absolute MUST Read IPCC Report: Debate Over, Further Delay Fatal, Action Not Costly

November 17, 2007

In its definitive scientific synthesis report, the Intergovernmental Panel on Climate Change (IPCC) today issued its strongest call for immediate action to save humanity from the deadly consequences of unrestrained greenhouse gas emissions.

This report—signed by more than 100 nations including the United States and China—slams the door on any argument for delay and makes clear we must under no circumstances listen to those who urge that we wait (who knows how long) to develop as yet nonexistent technology.* As the *New York Times* put it:

> Members of the panel said their review of the data led them to conclude as a group and individually that reductions in greenhouse gasses had to start immediately to avert a global climate disaster that could leave island states submerged and abandoned, African crop yields decreased by 50 percent, and cause over a 5 percent decrease in global gross domestic product.
>
> This summary was the first to acknowledge that the melting of the Greenland ice sheet from rising temperature [which would raise the oceans 23 feet] could result in sea-level rise over centuries rather than millennia.

And readers of this blog know the IPCC almost certainly underestimates the timing and severity of likely impacts because it ignores or downplays key amplifying feedbacks in the carbon cycle (discussed later in this chapter). Indeed, IPCC head Rajendra Pachauri—an engineer and economist—admitted as much to the *NY Times*:

* This means you President Bush, Newt Gingrich, Bjørn Lomborg.

> He said that since the panel began its work five years ago, scientists have recorded "much stronger trends in climate change," like a recent melting of polar ice that had not been predicted. "That means you better start with intervention much earlier."

How much earlier? The normally understated Pachauri warns:

> If there's no action before 2012, that's too late. What we do in the next two to three years will determine our future. This is the defining moment.

In short—time's up! *America—we better pick the right president in 2008.*

To balance the bad news, the IPCC and its member governments agree on the good news—action is very affordable, the report concludes:

> In 2050, global average macro-economic costs for mitigation towards stabilisation between 710 and 445ppm CO_2-eq are between a 1% gain and 5.5% decrease of global GDP. This corresponds to slowing average annual global GDP growth by less than 0.12 percentage points.

But how is that possible? How can the world's leading governments and scientific experts agree that we can avoid catastrophe for such a small cost?

Because that's what the scientific and economic literature—and real-world experience—says:

> Both bottom-up and top-down studies indicate that there is high agreement and much evidence of substantial economic potential for the mitigation of global GHG emissions over the coming decades that could offset the projected growth of global emissions or reduce emissions below current levels.

In fact, the bottom-up studies—the ones that look technology by technology, which I believe are more credible—have even better news:

> Bottom-up studies suggest that mitigation opportunities with *net negative costs* have the potential to reduce emissions by around 6 $GtCO_2$-eq/yr in 2030.

Wow! A 20 percent reduction in global emissions might be possible in a quarter century with net economic benefits! Take that, you delayers who oppose rapid, mandatory action and supposedly represent the "pragmatic center on climate and energy" (as the *New York Times* described them)—but who in fact represent the fatal siren song of "wait for new technology, wait for new technology."

But don't we need new technologies? Of course, but we don't need—and can't afford—to sit on our hands when we have so many cost-effective existing technologies:

> There is high agreement and much evidence that all stabilisation levels assessed can be achieved by deployment of a portfolio of technologies that are either currently available or expected to be commercialised in coming decades, assuming appropriate and effective incentives are in place for their development, acquisition, deployment, and diffusion and addressing related barriers.

Yes delayers—we need to do two things at once: aggressively deploy existing technology (with carbon prices and government standards) and aggressively finish developing and commercializing key technologies and systems that are in the pipeline. Anyone who argues for just doing the latter is disputing a very broad consensus—and is neither pragmatic nor centrist.

What do we risk if we fail to act now?

> Anthropogenic warming could lead to some impacts that are abrupt or irreversible, depending upon the rate and magnitude of the climate change.

Partial loss of ice sheets on polar land could imply meters of sea level rise, major changes in coastlines, and inundation of low-lying areas, with greatest effects in river deltas and low-lying islands. Such changes are projected to occur over millennial time scales, but more rapid sea level rise on century time scales cannot be excluded.

In short, we risk that our top climatologists' warnings on sea level rise prove true. Also, "As global average temperature increase exceeds about 3.5°C, model projections suggest significant extinctions (40–70% of species assessed) around the globe."

IPCC to world: The time to act is now or we risk destroying life on the Earth as we know it today!

· An Introduction to Global Warming Impacts: Hell and High Water

March 22, 2009

In this post, I will examine the key impacts we face by 2100 if we stay anywhere near our current emissions path. I will focus primarily on:

- Staggeringly high temperature rise, especially over land—some 10°F over much of the United States
- Sea level rise of some 5 feet, rising several inches (or more) each decade thereafter
- Dust Bowls over the southwestern United States and many other heavily populated regions around the globe
- Massive species loss on land and sea—50 percent or more of all life
- Unexpected impacts—the fearsome "unknown unknowns"

Equally tragic, a 2009 NOAA-led study found the worst impacts would be "largely irreversible for 1000 years."

The single biggest failure of messaging by climate scientists (until very recently) has been the failure to explain to the public, opinion makers, and the media that business-as-usual warming results in impacts that are beyond catastrophic. For these impacts, terms like "global warming" and "climate change" are essentially euphemisms.

Business-as-usual typically means continuing at recent growth rates of carbon dioxide emissions, which we now know would take us to atmospheric concentrations of carbon dioxide greater than 1000 ppm (parts per million). We are at nearly 390 ppm today, up from pre-industrial levels of about 275 ppm.

The scientific community has spent little time modeling the impacts of tripling (~830 ppm) or quadrupling (~1100 ppm) carbon dioxide concentrations from preindustrial levels. In part, I think, that's because they never believed humanity would be so stupid as to ignore their warnings and simply continue on its self-destructive path. In part, they lowballed the difficult-to-model amplifying feedbacks in the carbon cycle.

So I pieced together those impacts from available studies and from discussions with leading climate scientists for my book, *Hell and High Water.* But now as climate scientists have sobered up to their painful role as modern-day Cassandras, the scientific literature on what we face is much richer. Let me review it here.

Temperature

Two of the best recent analyses of what we are headed toward can be found here:

- MIT joins climate realists, doubles its projection of global warming by 2100 to 5.1°C (discussed later in this chapter)
- Hadley Centre: "Catastrophic" 5-7°C warming by 2100 on current emissions path

Dr. Vicky Pope, head of Climate Change Advice for the UK Met Office's Hadley Centre, explains the second analysis on their website:

> Contrast that with a world where no action is taken to curb global warming. Then, temperatures are likely to rise by 5.5°C and could rise as high as 7°C above preindustrial values by the end of the century.

That likely rise corresponds to roughly 9°F globally and typically much higher than that over inland mid-latitudes (i.e., much of this country)—as high as 12°F.

Based on two studies in the last few years:

> By century's end, extreme temperatures of up to 122°F would threaten most of the central, southern, and western United States. Even worse, Houston and Washington, DC, could experience temperatures exceeding 98°F for some 60 days a year. Much of Arizona would be subjected to temperatures of 105°F or more for 98 days out of the year—14 full weeks.

Yet that conclusion is based on studies of *only* 700 ppm and 850 ppm, so it could get much hotter than that.

The Hadley Centre adds, "By the 2090s, close to one-fifth of the world's population will be exposed to ozone levels well above the World Health Organization recommended safe-health level."

The Hadley Centre has a huge but useful figure that I reproduce here:

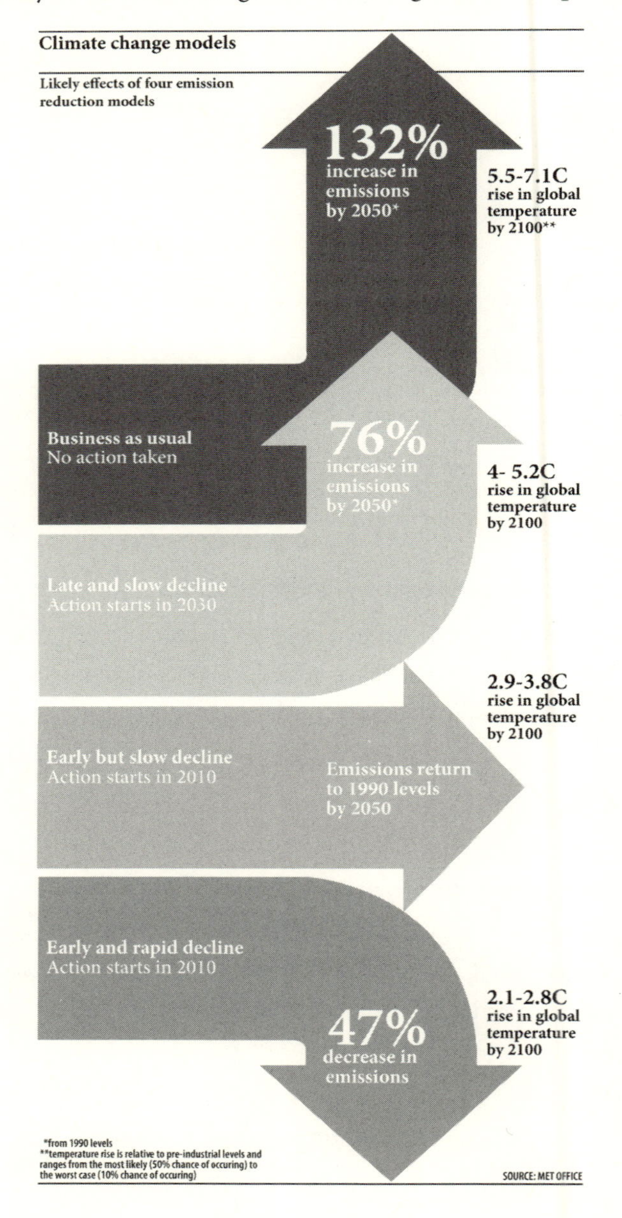

Sea Level Rise

A 5.5°C (9.9°F) warming would likely lead to the mid- to high-range of currently projected sea level rise (SLR)—5 feet or more by 2100, followed by perhaps 6 to 12 inches a decade for centuries. One of the most comprehensive recent studies is "Kinematic Constraints on Glacier Contributions to 21st-Century Sea-Level Rise," which concludes:

> On the basis of calculations presented here, we suggest that an improved estimate of the range of SLR to 2100 including increased ice dynamics lies between 0.8 and 2.0 m.
>
> These values give a context and starting point for refinements in SLR forecasts on the basis of clearly defined assumptions and offer a more plausible range of estimates than those neglecting the dominant ice dynamics term.

Scientific analysis is finally catching up to scientific observation. In 2001, the IPCC projected that neither Greenland nor Antarctica would lose significant mass by 2100. The IPCC made the same basic projection again in 2007. Yet both ice sheets *already* are losing mass. As Penn State climatologist Richard Alley said in March 2006, the ice sheets appear to be shrinking "100 years ahead of schedule."

Science's Richard Kerr explained what the IPCC did wrong and what the new study does right:

> Warming glaciers raise sea level in two main ways. They add more water as they melt, and they also add water when ice breaks off from glacial flows. The incidence of this latter phenomenon has soared in recent years for some glaciers draining the southern Greenland Ice Sheet, much to the mystification of glaciologists. Unable to model such accelerated ice losses, members of the Intergovernmental Panel on Climate Change declined to include them in their widely cited projection of up to 60 centimeters of sea level rise by 2100.
>
> Glaciologist W. Tad Pfeffer of the University of Colorado, Boulder, and his colleagues tackled glacier flow anyway. They calculated how fast glaciers would have to flow in order to raise sea level by a given number of meters and then considered whether those flow rates were plausible or even physically possible.

Needless to say, a sea level rise of 1 meter by 2100 would be an unmitigated catastrophe for the planet, even if sea levels didn't keep rising several inches a decade for centuries, which they inevitably would. The first meter of SLR would flood 17 percent of Bangladesh, displacing tens of millions of people, and reducing its rice-farming land by 50 percent. Globally, it would create more than 100 million environmental refugees and inundate more than 13,000 square miles of the United States. Southern Louisiana and South Florida would inevitably be abandoned. And salt water infiltration will only compound this impact. As will stronger storm surges.

The scientific literature has been moving in this direction for a couple of years now—too late for the IPCC to consider in its latest assessment. For instance, an important *Science* article from 2007 used empirical data from last century to project that sea levels could be up to 5 feet higher in 2100 and rising 6 inches a decade.

Another 2007 study from *Nature Geoscience* came to the same conclusion (see my ClimateProgress.org post "Sea Levels May Rise 5 Feet by 2100"). Leading experts in the field have a similar view (see my post, "Report from AGU Meeting: One Meter Sea Level Rise by 2100 'Very Likely' Even If Warming Stops?").

The anti-science disinformers like to hide behind the IPCC's 2007 sea level estimate—even though they really don't believe most of what the IPCC says or most of the scientific literature on which it bases its conclusion. So, you're going to be hearing the IPCC estimate for another several years, until the IPCC does a new report and puts in a more realistic estimate. That said, while the disinformers never acknowledge it, even the 2007 IPCC report "was the first to acknowledge that the melting of the Greenland ice sheet from rising temperature [which would raise the oceans 23 feet] could result in sea-level rise over centuries rather than millennia," as the *New York Times* put it.

Even a major report signed off on by the Bush administration itself was forced to concede that the IPCC numbers are simply too out of date to be quoted anymore—see my blog post, "U.S. Geological Survey Stunner: Sea Level Rise in 2100 Will Likely 'Substantially Exceed' IPCC Projections."

Dust-Bowlification

Then we have moderate drought over half the planet, plus the loss of all inland glaciers that provide water to a billion people.

"The unexpectedly rapid expansion of the tropical belt constitutes yet another signal that climate change is occurring sooner than expected," noted one climate researcher in December 2007. A 2008 study led by NOAA noted, "A poleward expansion of the tropics is likely to bring even drier conditions to the U.S. Southwest, Mexico, Australia, and parts of Africa and South America."

In 2007, *Science* published research that "predicted a permanent drought by 2050 throughout the Southwest"—levels of aridity comparable to the 1930s Dust Bowl would stretch from Kansas to California. And they were only looking at a 720 ppm case! The Dust Bowl was a sustained decrease in soil moisture of about 15 percent ("which is calculated by subtracting evaporation from precipitation").

A NOAA-led study similarly found permanent Dust Bowls in Southwest and around the globe on our current emissions trajectory (and irreversibly so for 1,000 years). And future droughts will be fundamentally different from previous droughts that humanity has experienced because they will be very hot weather droughts.

I should note that even the "moderate drought over half the planet" scenario from the Hadley Centre is *only* based on 850 ppm (in 2100). Princeton has published an analysis titled "Century-Scale Change in Water Availability: CO_2-Quadrupling Experiment," which is to say 1100 ppm. The grim result: Most of the South and Southwest will ultimately see a 20 percent to 50 percent (!) decline in soil moisture.

Species Loss on Land and Sea

In 2007, the IPCC warned that "as global average temperature increase exceeds about 3.5°C [relative to 1980 to 1999], model projections suggest significant extinctions (40–70% of species assessed) around the globe." That is a temperature rise over preindustrial levels of a bit more than 4.0°C. So a 5.5°C rise would likely put extinctions beyond the high end of that range.

And, of course, "when CO_2 levels in the atmosphere reach about 500 parts per million, you put calcification [crucial for corals and other marine life] out of business in the oceans." There aren't many studies of what happens to the oceans as we get toward 800 to 1000 ppm, but it appears likely that much of the world's oceans, especially in the southern hemisphere, become inhospitable to many forms of marine life because of acidification. A 2005 *Nature* study concluded that these "detrimental" conditions "could develop within decades, not centuries as suggested previously."

A 2009 study in *Nature Geoscience* warned that global warming may create "dead zones" in the ocean that would be devoid of fish and seafood and endure for up to two millennia.

Unexpected Impacts

If we go to 800 ppm—let alone 1000 ppm or higher—we are far outside the bounds of simple linear projection. Some of the worst impacts may not be obvious—and there may be unexpected negative synergies. The best evidence that will happen is the fact that it has already happened with even the small amount of warming we have seen to date.

"The pine beetle infestation is the first major climate change crisis in Canada" notes Doug McArthur, a professor at Simon Fraser University in Vancouver. The pests are "projected to kill 80 percent of merchantable and susceptible lodgepole pine" in parts of British Columbia within ten years—and that's why the harvest levels in the region have been "increased significantly."

As quantified in the journal *Nature*, "Mountain Pine Beetle and Forest Carbon Feedback to Climate Change," which just looks at the current and future impact from the beetle's warming-driven devastation in British Columbia:

> The cumulative impact of the beetle outbreak in the affected region during 2000–2020 will be 270 megatonnes (Mt) carbon (or 36 g

> carbon m^{-2} yr^{-1} on average over 374,000 km^2 of forest). This impact converted the forest from a small net carbon sink to a large net carbon source.

No wonder some carbon sinks are losing their ability to absorb carbon dioxide faster than we thought—unmodeled impacts of climate change are destroying them:

> Insect outbreaks such as this represent an important mechanism by which climate change may undermine the ability of northern forests to take up and store atmospheric carbon, and such impacts should be accounted for in large-scale modeling analyses.

And the bark beetle is slamming the Western U.S. and Alaska, too (see my blog post, "Oldest Utah Newspaper: Bark-Beetle Driven Wildfires Are a Vicious Climate Cycle").

The key point is that this catastrophic climate change impact and its carbon-cycle feedback were not foreseen even a decade ago—which suggests future climate impacts will bring other equally unpleasant surprises, especially as we continue on our path of no action.

Conclusion

We can't let this happen. We must pay any price or bear any burden to stop it.

And let me make one final point. I think it is increasingly clear the "middle ground" scenarios are unstable in that once you hit 500 ppm (or possibly lower), the amplifying feedbacks kick in. As Dr. Pope puts it, "If the climate turns out to be particularly sensitive to increases in greenhouse gases and the Earth's biological systems cannot absorb very much carbon then temperature rises could be even higher."

Indeed, some of the best research on this has come from the Hadley Centre, since it has one of the few models that incorporate many of the major carbon cycle feedbacks. In a 2003 *Geophysical Research Letters* paper, "Strong carbon cycle feedbacks in a climate model with interactive CO_2 and sulphate aerosols," the Hadley Centre, the UK's official center for climate change research and part of their Defence Ministry, finds that the world would hit 1000 ppm in 2100 even in a scenario that, absent those feedbacks, we would have only hit 700 ppm in 2100. I would note that the Hadley Centre, though more inclusive of carbon cycle feedbacks than most other models, still does not model most feedbacks, including *any* feedbacks from the carbon dioxide and methane release from the melting of the tundra and permafrost, even though that is probably the most serious of the amplifying feedbacks.

So we must stabilize at 450 ppm or below—or risk what can only be called humanity's self-destruction. Since the cost is maybe 0.12 percent of GDP per year—or probably a bit higher than that if we shoot for 350 ppm—the choice is clear. Now if only the scientific community and environmentalists and progressives could start articulating this reality cogently.

U.S. Media Largely Ignores Latest Warning from Climate Scientists: "Recent Observations Confirm . . . the Worst-Case IPCC Scenario Trajectories (Or Even Worse) Are Being Realised"—1000 ppm

March 17, 2009

In the last two years, our scientific understanding of business-as-usual projections for global warming has changed dramatically. Yet, much of the U.S. public—especially conservatives—remain in the dark about just how dire the situation is.

Why? Because the U.S. media is largely ignoring the story. Case in point: Where was the coverage of the Copenhagen Climate Science Congress, attended by 2,000 scientists, which concluded with this Key Message #1:

> Recent observations confirm that, given high rates of observed emissions, the worst-case IPCC scenario trajectories (or even worse) are being realized. For many key parameters, the climate system is already moving beyond the patterns of natural variability within which our society and economy have developed and thrived. These parameters include global mean surface temperature, sea level rise, ocean and ice sheet dynamics, ocean acidification, and extreme climatic events. There is a significant risk that many of the trends will accelerate, leading to an increasing risk of abrupt or irreversible climatic shifts.

What is the worst-case IPCC scenario trajectory? That would be A1F1 (the red dotted line in figure 2.2 from figure SPM-3 of the 2001 Intergovernmental Panel on Climate Change, Synthesis Report):

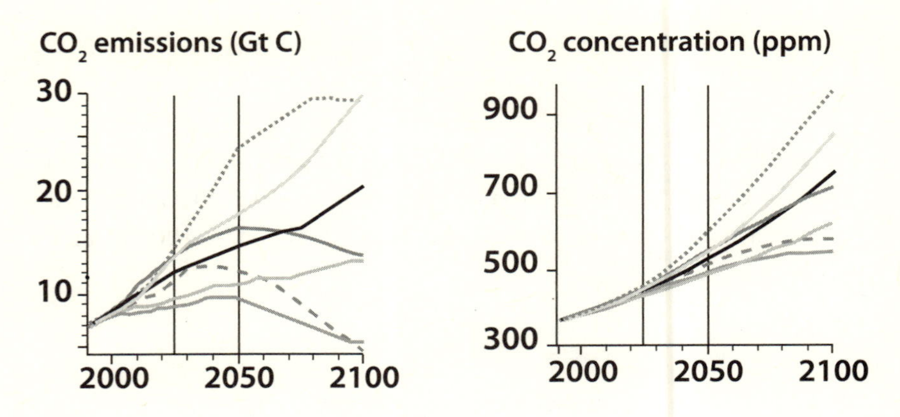

The A1F1 scenario takes us to atmospheric concentrations of carbon dioxide of 1000 ppm in 2100—otherwise known as the end of human civilization as we have known it.

Actually it's worse than that. The 2001 IPCC report largely failed to model amplifying carbon cycle feedbacks. The 2007 IPCC report, which began to consider such feedbacks, warns that even averaging 11 GtC (billion metric tons of carbon) a year this century could take us to 1000 ppm. The A1F1 scenario averages well above 15 GtC a year through 2100 as you can see from the figure on the left.

Energy Daily notes of the U.S. media non-coverage of Copenhagen:

> Ironically—given the Gallup finding that two in five Americans think the press is exaggerating climate change concerns—only a few of the major U.S. news outlets published accounts of the Copenhagen gathering, which received heavy coverage by news outlets in Europe and Asia.

Great point—though "ironically" isn't the right word. There is nothing ironic about this. It is cause and effect. The right word is "tragically."

Exceeding A1F1 probably means total planetary warming by 2100 compared to preindustrial levels of 5°C or more. I discuss the harsh impacts of such warming above.

West Coast Climate Equity notes:

> Last time mean global temperatures reached 2 to 3 degrees Celsius above present levels, in the mid-Pliocene (3 million years ago), an event associated with CO2 levels of about 400 parts per million, polar regions were

> heated by near-8°C and sea levels have risen by 25+/–12 meters relative to the present. This represents near-total melting of Greenland and west Antarctica ice sheets. . . .
>
> A rise of mean global temperatures above 4 or 5 degrees Celsius would shift the atmosphere to pre-glacial/interglacial conditions, which dominated the Earth from about 34 million years ago.

That means ultimate sea level rise of 250 feet, with the best current projection being 5 feet by 2100, rising thereafter 6 to 12 inches a decade (or more) for centuries. Good luck adapting to that, future generations.

Key Message #5 from the Congress is:

> There is no excuse for inaction. We already have many tools and approaches—economic, technological, behavioural, management—to deal effectively with the climate change challenge. But they must be vigorously and widely implemented to achieve the societal transformation required to decarbonise economies. A wide range of benefits will flow from a concerted effort to alter our energy economy now, including sustainable energy job growth, reductions in the health and economic costs of climate change, and the restoration of ecosystems and revitalisation of ecosystem services.

What is inexcusable is U.S. media coverage and the blinkered conservative strategy of scientific denial—what can only be described as a murder-suicide pact with the human race.

MIT Joins Climate Realists, Doubles Its Projection of Global Warming by 2100 to 5.1°C

February 23, 2009

The Massachusetts Institute of Technology Joint Program on the Science and Policy of Climate Change has joined the climate realists. The realists are the growing group of scientists who understand that the business as usual emissions path leads to unmitigated catastrophe (such as the Hadley Centre, which warns of "catastrophic" 5 to 7°C warming by 2100 on the current emissions path).

The Program issued a remarkable, though little-remarked-on, report in January, "Probabilistic Forecast for 21st Century Climate Based on Uncertainties in Emissions (without Policy) and Climate Parameters," by over a dozen leading experts. They reanalyzed their model's 2003 projections model using the latest data, and concluded:

> The MIT Integrated Global System Model is used to make probabilistic projections of climate change from 1861 to 2100. Since the model's first projections were published in 2003 substantial improvements have been made to the model and improved estimates of the probability distributions of uncertain input parameters have become available. The new projections are considerably warmer than the 2003 projections, e.g., the median surface warming in 2091 to 2100 is 5.1°C compared to 2.4°C in the earlier study.†

Their median projection for the atmospheric concentration of carbon dioxide in 2095 is a jaw-dropping 866 ppm.

---

† That rise is compared to 1990 levels. So you can add at least 0.5°C and 1.0°F for comparison with preindustrial temperatures.

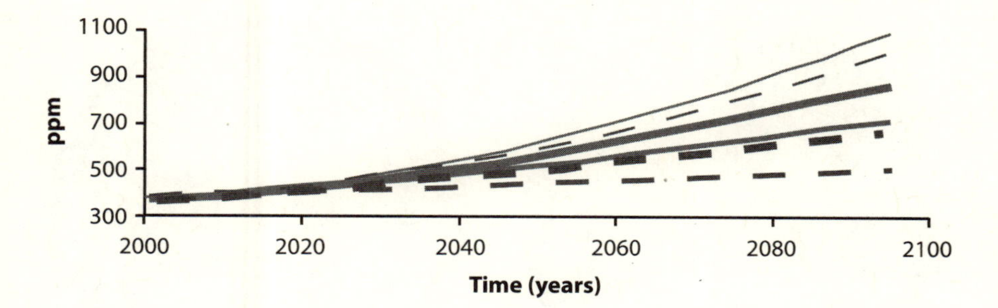

Why the change? The Program's website explains:

> There is no single revision that is responsible for this change. In our more recent global model simulations, the ocean heat-uptake is slower than previously estimated, the ocean uptake of carbon is weaker, feedbacks from the land system as temperature rises are stronger, cumulative emissions of greenhouse gases over the century are higher, and offsetting cooling from aerosol emissions is lower. No one of these effects is very strong on its own, and even adding each separately together would not fully explain the higher temperatures. Rather than interacting additively, these different affects appear to interact multiplicatively, with feedbacks among the contributing factors, leading to the surprisingly large increase in the chance of much higher temperatures.

The carbon sinks are saturating, and the amplifying feedbacks are worse than previously thought—that, of course, is a central understanding of all climate realists.

Andrew Freedman at Washingtonpost.com has one of the very few stories on this important study and reprints this useful figure from MIT:

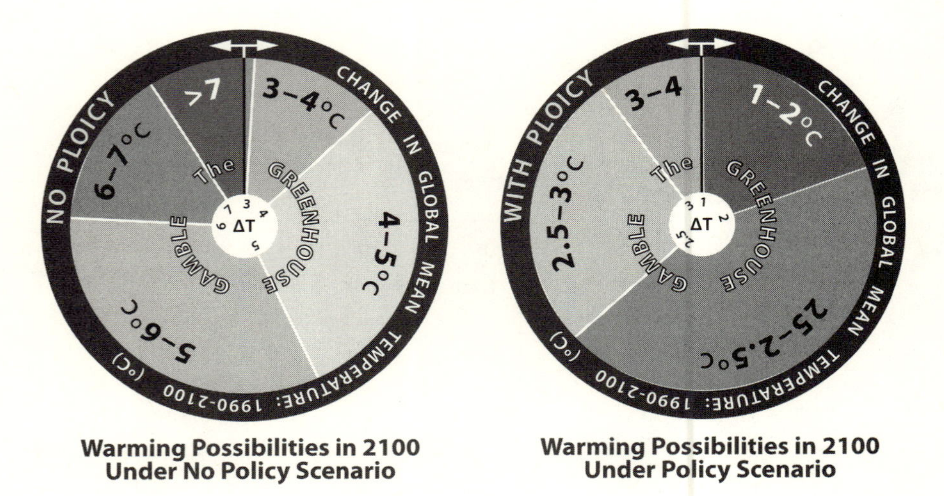

Warming Possibilities in 2100
Under No Policy Scenario

Warming Possibilities in 2100
Under Policy Scenario

He explains:

> Results of the studies are depicted online in MIT's "Greenhouse Gamble" exercise that conveys the "range of probability of potential global warming" via roulette wheel graphics (shown above). The modeling output showed that under both a "no policy" scenario and one in which nations took action beginning in the next few years to reduce greenhouse gas emissions, the odds have shifted in favor of larger temperature increases.
>
> For the no-policy scenario, the researchers concluded that there is now a 9 percent chance (about one in 11 odds) that the global average surface temperature would increase by more than 7°C (12.6°F) by the end of this century, compared with only a less than 1 percent chance (one in 100 odds) that warming would be limited to below 3°C (5.4°F).

To repeat, on our current emissions path, we have a 9 percent chance of an incomprehensibly catastrophic warming of 7°C by century's end, but less than a 1 percent chance of under 3°C warming.

The model shows staggering warming near the poles. Median arctic warming—north of 70° latitude—is 20°F! How could Greenland's ice sheet possibly survive that?

> "The take home message from the new greenhouse gamble wheels is that if we do little or nothing about lowering greenhouse gas emissions that the dangers are much greater than we thought three or four years ago,"

> said Ronald G. Prinn, professor of atmospheric chemistry at MIT. "It is making the impetus for serious policy much more urgent than we previously thought."

The time to act is now.

Study: Water-Vapor Feedback Is "Strong and Positive," So We Face "Warming of Several Degrees Celsius"

October 26, 2008

A new study in *Geophysical Research Letters*, "Water-vapor Climate Feedback Inferred from Climate Fluctuations, 2003–2008," analyzed recent variations in surface temperature and "the response of tropospheric water vapor to these variations." They concluded that the "water-vapor feedback implied by these observations is strongly positive" and "similar to that simulated by climate models." The analysis concludes:

> The existence of a strong and positive water-vapor feedback means that projected business-as-usual greenhouse-gas emissions over the next century are virtually guaranteed to produce warming of several degrees Celsius. The only way that will not happen is if a strong, negative, and currently unknown feedback is discovered somewhere in our climate system.

A "warming of several degrees Celsius" is the end of civilization as we know it.

While some disinformers/delayers/inactivists, like MIT's Richard Lindzen, have argued that negative feedbacks dominate the climate—all of the evidence points to amplifying feedbacks dominating (except the one negative feedback that the deniers fiercely fight, discussed below).

In the real world, key climate change impacts—sea ice loss, ice sheet melting, desertification, and sea level rise—all are either near the top or actually in excess of their values as predicted by the IPCC's climate models.

The major climate models are missing key amplifying feedbacks, whereby a small initial warming leads to a disproportionately huge heating:

- The defrosting of the permafrost
- The drying of the Northern peatlands (bogs, moors, and mires)

- The destruction of the tropical wetlands
- Decelerating growth in tropical forest trees—thanks to accelerating carbon dioxide
- Wildfires and climate-driven forest destruction by pests
- The desertification–global warming feedback
- The saturation of the ocean carbon sink

And this all supports the analysis that the climate is much more sensitive to changes in greenhouse gas emissions and other "forcings" than the IPCC models have been saying and that a doubling of atmospheric concentrations of carbon dioxide from preindustrial levels to 550 ppm will ultimately warm the planet far more than 3°C. NASA's James Hansen argues the "long-term" climate sensitivity for doubled CO_2 is of 6°C (10°F).

A number of major studies looking at paleoclimate data come to the same conclusion. Here are three:

Scientists analyzed data from a major expedition to retrieve deep marine sediments beneath the Arctic to understand the Paleocene Eocene thermal maximum, a brief period some 55 million years ago of "widespread, extreme climatic warming that was associated with massive atmospheric greenhouse gas input." A 2006 study, published in *Nature*, found Artic temperatures almost beyond imagination—*above 23°C (74°F)*—temperatures more than 18°F warmer than current climate models had predicted when applied to this period. The three dozen authors conclude that existing climate models are missing crucial feedbacks that can significantly amplify polar warming.

A second study, published in *Geophysical Research Letters*, looked at temperature and atmospheric changes during the Middle Ages. This 2006 study found that the effect of amplifying feedbacks in the climate system—where global warming boosts atmospheric CO_2 levels—"will promote warming by an extra 15 percent to 78 percent on a century-scale" compared to typical estimates by the UN's Intergovernmental Panel on Climate Change. The study notes these results may even be "conservative" because they ignore other greenhouse gases such as methane, whose levels will likely be boosted as temperatures warm.

The third study, published in *Geophysical Research Letters*, looked at temperature and atmospheric changes during the last 400,000 years. This study found evidence for significant increases in both CO_2 and methane (CH_4) levels as temperatures rise. The conclusion: If our current climate models correctly accounted for such "missing feedbacks," then "we would be predicting a significantly greater increase in global warming than is currently forecast over the next century and beyond"—as much as 1.5°C warmer this century alone.

Yes, natural negative feedbacks exist that would "eventually" absorb any excess carbon dioxide, but as one of the authors of a 2008 *Nature Geosciences* article explained, "not for hundreds of thousands of years." Humans are boosting carbon dioxide levels 14,000 times faster than nature, overwhelming slow negative feedbacks.

Truly only one negative feedback in the planet's overall carbon cycle can act with sufficient speed and strength to avert catastrophic climate impacts: The dominant carbon-based life form on this planet will have to respond to the already painfully clear impacts of our carbon emissions by slashing those emissions sharply and eventually running the planet on carbon-negative power.

The time for this negative feedback is now.

CHAPTER 3

The Clean Energy Solution

IN 2005, BRITISH PRIME MINISTER Tony Blair described the crucial two-pronged strategy we must adopt to preserve a livable climate: "We need to invest on a large scale in existing technologies *and* to stimulate innovation into new low-carbon technologies for deployment in the longer term." Future technology will be able to help preserve our way of life in the long term *if and only if* we have already moved "on a large scale" to technologies that already exist.

Those who don't believe global warming is serious—the climate science disinformers and delayers, and those who have been duped by them—invariably say we need a massive investment in research and development into new technologies to solve this problem at an affordable price. They don't actually support such an investment, and in fact conservatives have historically opposed all efforts to ramp up clean energy R&D, starting with Ronald Reagan, who slashed federal spending on renewable energy in the early 1980s.

For several years in the mid-1990s, I helped manage the Energy Department's billion-dollar Office of Energy Efficiency and Renewable Energy. It was the largest program in the world for working with businesses to develop, demonstrate, and deploy clean energy and low-carbon technologies.

My conclusion is that we have all the technology we need to start reducing emissions aggressively and cost-effectively now—particularly core solutions like energy efficiency and the under-discussed concentrated solar thermal power, "the technology that will save humanity." And as we'll see the three best strategies for lowering the cost of clean energy technologies are deployment, deployment, and deployment.

Moreover, the rate of technological change is happening so fast the only way to stay up with the dramatic changes is online. Just since I've started writing this blog, for instance, the global wind industry has doubled in size, and the United States has regained leadership in wind power deployment. Over the same time period, concentrated solar power has gone from a fledgling industry with few players and fewer contracts to one of the fastest-growing forms of low-carbon power, with huge international companies entering the business and more than 10,000 MW of global contracts.

How the World Can (and Will) Stabilize at 350 to 450 ppm: The Full Global Warming Solution

March 26, 2009

Stabilizing atmospheric concentrations of carbon dioxide at 450 ppm or lower may not be politically possible today, but it is certainly achievable from an economic and technological perspective. I do, however, believe humanity will do it since the alternative is Hell and high water.

It would require some twelve to fourteen of Princeton's "stabilization wedges":

> A wedge represents an activity that reduces emissions to the atmosphere that starts at zero today and increases linearly until it accounts for 1 GtC/year [1 billion metric tons per year] of reduced carbon emissions in 50 years. It thus represents a cumulative total of 25 GtC of reduced emissions over 50 years.

We need twice as many wedges as Princeton's Pacala and Socolow have said because their 2004 paper is now quite out of date. That conclusion is also corroborated by the 2008 International Energy Agency (IEA) report, *Energy Technology Perspectives, 2008*, which found stabilization at 450 ppm can be achieved with the equivalent of about thirteen wedges starting by 2015.

I agree with the IPCC's detailed review of the technical literature, which concluded in 2007 that "The range of stabilization levels assessed can be achieved by deployment of a portfolio of technologies that are currently available and those that are expected to be commercialised in coming decades." *Technology Review*, one of the nation's leading technology magazines, also argued in a 2006 cover story, "It's Not Too Late," that "catastrophic climate change is not inevitable. We possess the technologies that could forestall global warming."

Pacala and Socolow made the same key point in their analysis: "Humanity already possesses the fundamental scientific, technical, and industrial know-how to solve the carbon and climate problem for the next half-century."

I do believe only "one" solution exists in this sense: We must deploy every conceivable energy-efficient and low-carbon technology that we have today as fast as we can. Princeton's Pacala and Socolow proposed that this could be done over fifty years, but that is almost certainly too slow. The longer we delay acting, the faster we need to deploy technology.

We're at about 30 billion tons of carbon dioxide emissions a year—and notwithstanding the global economic slowdown, probably poised to rise 2 percent per year (the exact future growth rate is quite hard to project because it depends so much on what China does and how quickly peak oil kicks in). We have to average below 18 billion tons (below 5 GtC) a year for the entire century if we're going to stabilize at 450 ppm. As 200 of our top climate scientists said in 2007, we need to peak around 2020, then drop at least 60 percent by 2050 to 15 billion tons at most (4 billion tons of carbon), and then go to near zero net carbon emissions by 2100.

If we could do the twelve to fourteen wedges in four decades, we should be able to keep CO_2 concentrations to under 450 ppm. If we could do them faster, concentrations could stay even lower. We'd probably need to do this by 2040 if not sooner to have a shot at getting back to 350 ppm this century. And, yes, like Princeton, I agree we need to do some R&D now to ensure a steady flow of technologies to make the even deeper emissions reductions needed in the second half of the century.

I do agree with James Hansen et al. that the basic strategy is to replace virtually all of coal as quickly as possible, which is why so many of the wedges focused on electricity—that, along with the need to electrify transportation as much as possible. I also agree that this will be harder and more expensive if conventional oil were not going to peak soon. But for better or worse, it is (see the next chapter).

Also, I tend to view the crucial next four decades in two phases. In Phase

1, 2010 to 2030, the world finally gets serious about avoiding catastrophic global warming impacts. We increasingly embrace a serious price for carbon dioxide and a very aggressive technology deployment effort.

In Phase 2, 2030 to 2050, after the visceral reality of climate change is obvious to even the most conservative denier, after the inevitable collapse of the Ponzi scheme we call the global economy, the world gets truly desperate, and actions that are not plausible today—including widespread conservation—become commonplace.

In the basic solution below, I have thrown in some extra wedges since, no doubt, everybody will find something objectionable in at least two of them. You can find almost all of these solutions discussed at length on ClimateProgress.org. Three core solutions—efficiency, concentrated solar, and electric cars—are discussed in this chapter and the next.

This is what the entire planet must achieve:

- One wedge of albedo change through white roofs and pavement (i.e., to reflect more sunlight).
- One wedge of vehicle efficiency—all cars 60 mpg, with no increase in miles traveled per vehicle.
- One of wind for power—1 million large (2 MW peak) wind turbines.
- One of wind for vehicles—another 2000 GW wind. Most cars must be plug-in hybrids or pure electric vehicles.
- Three of concentrated solar thermal—5000 GW peak.
- Three of efficiency—one each for buildings, industry, and cogeneration/heat-recovery for a total of 15 to 20 million GW-hrs. A key strategy for reducing direct fossil fuel use for heating buildings (while also reducing air conditioning energy) is geothermal heat pumps.
- One of solar photovoltaics—2000 GW peak.
- One half wedge of nuclear power—350 GW.
- Two of forestry—end all tropical deforestation. Plant new trees over an area the size of the continental United States.
- One wedge of WWII-style conservation, post-2030.

Here are additional wedges that require some major advances in applied research to be practical and scalable, but are considered plausible by serious analysts, especially post-2030:

- One of geothermal power plus ocean-based renewables (i.e., tidal, wave, and/or ocean thermal).
- One of coal with biomass cofiring plus carbon capture and storage—400 GW of coal plus 200 GW biomass with CCS.
- One half wedge of next-generation nuclear power—350 GW.
- One half wedge of cellulosic biofuels for long-distance transport and what's left of aviation in 2050—using 8 percent of the world's cropland (or less land if yields significantly increase or algae-to-biofuels proves commercial at large scale).
- One of soils and/or biochar—apply improved agricultural practices to all existing croplands and/or "charcoal created by pyrolysis of biomass." Both are controversial today, but may prove scalable strategies.

That should do the trick. And, yes, the scale is staggering.*

Why not more than one wedge of CCS? That one wedge represents a flow of CO_2 into the ground equal to the current flow of oil out of the ground. It would require, by itself, recreating the equivalent of the planet's entire oil delivery infrastructure. I also think that CCS has practical issues that will limit its scale, not the least of which is that I doubt it will be among the cheaper solutions. But the possibility of doing coal and biomass gasification together with CCS—resulting in negative-carbon electricity that actually pulls CO_2 out of the air—makes this too important a strategy not to pursue aggressively.

Why not more than one total wedge of nuclear? To do this by 2050 would require adding, globally, an average of seventeen plants each year, while building an average of nine plants a year to replace those that will be retired, for a total of one nuclear plant every two weeks for four decades—plus ten Yucca Mountains to store the waste. I also doubt it will be among the cheaper options. And the uranium supply and nonproliferation issues for even that scale of deployment are quite serious.

Do I want to build all those nuclear plants? No. Do I think we could do it without all those nuclear plants? Definitely. Therefore, should I be quoted

---

* For those who prefer terawatts, 1000 GW=1 TW. I have adjusted the peak GW of the renewable wedges to take into account the lower capacity factor of solar and wind. The efficiency measures are assumed to have a capacity factor of about 60 percent.

as saying we "must" build all those nuclear plants, as the Drudge Report has, or even that I *propose* building all those plants? No. Do I think we will have to swallow a bunch of nuclear plants as part of the grand bargain to make this all possible and that other countries will build most of these? I have no doubt. So it stays in "the solution" for now.†

This is not to say the two wind power wedges (4000 GW peak total) would be easy—but the world did build more than 27 GW in 2008, a 36 percent jump from 2007. We would need to average 100 GW/year through 2050. But I do think it is ecologically and economically possible, as I think all the other wedges in the top group are, too.

But none of the wedges is easy. That's why getting to 450 ppm is not yet politically possible. Not even close.

Three more points: First, it bears repeating that the wedges are not analytically rigorous, but they are conceptually useful. We might need a couple more or a couple fewer.

Second, some people mistakenly think we need a lot more wedges. I have explained where they are wrong on this blog.

Third, if you don't like one of those wedges, you need to find a replacement strategy. Other possibilities can be found in the original paper by Pacala and Socolow, but I think the ones above are the most plausible by far, which tells you how dubious some of Princeton's other wedges are—*I'm talking about you, would-be hydrogen wedges.*

Could a bunch of breakthrough technologies substitute for some of the above wedges? That is far, far more implausible, as I discuss later.

---

† Note to self: Are you beginning to sound like Donald Rumsfeld? Yes.

Why We Never Need to Build Another Polluting Power Plant: Coal? Natural Gas? Nuke? We Can Wipe Them All Off the Drawing Board by Using Current Energy More Efficiently.

July 28, 2008

Suppose I paid you for every pound of pollution you generated and punished you for every pound you reduced. You would probably spend most of your time trying to figure out how to generate more pollution. And suppose that if you generated enough pollution, I had to pay you to build a new plant, no matter what the cost, and no matter how much cheaper it might be to not pollute in the first place.

Well, that's pretty much how we have run the U.S. electric grid for nearly a century. The more electricity a utility sells, the more money it makes. If it's able to boost electricity demand enough, the utility is allowed to build a new power plant with a guaranteed profit. The only way a typical utility can lose money is if demand drops. So the last thing most utilities want to do is seriously push strategies that save energy, strategies that do not pollute in the first place.

America is the Saudi Arabia of energy waste. A 2007 report from the international consulting firm McKinsey and Co. found that improving energy efficiency in buildings, appliances, and factories could offset almost all of the projected demand for electricity in 2030 and largely negate the need for new coal-fired power plants. McKinsey estimates that one-third of the U.S. greenhouse gas reductions by 2030 could come from electricity efficiency and be achieved at negative marginal costs. In short, the cost of the efficient equipment would quickly pay for itself in energy savings.

While a few states have energy-efficiency strategies, none matches what California has done. In the past three decades, electricity consumption per capita grew 60 percent in the rest of the nation, while it stayed flat in high-tech, fast-growing California. If all Americans had the same per capita electricity demand as Californians currently do, we would cut electricity consumption 40 percent. If the entire nation had California's much cleaner electric grid, we would cut total U.S. global-warming pollution by more than a quarter without raising American electric bills. And if all of America adopted the same energy-efficiency policies that California is now putting in place, the country would never have to build another polluting power plant.

How did California do it? In part, a smart California Energy Commission has promoted strong building standards and the aggressive deployment of energy-efficient technologies and strategies—and has done so with support of both Democratic and Republican leadership over three decades.

Many of the strategies are obvious: better insulation, energy-efficient lighting, heating, and cooling. But some of the strategies were unexpected. The state found that the average residential air duct leaked 20 to 30 percent of the heated and cooled air it carried. It then required leakage rates below 6 percent, and every seventh new house is inspected. The state found that in outdoor lighting for parking lots and streets, about 15 percent of the light was directed up, illuminating nothing but the sky. The state required new outdoor lighting to cut that to below 6 percent. Flat roofs on commercial buildings must be white, which reflects the sunlight and keeps the buildings cooler, reducing air-conditioning energy demands. The state subsidized high-efficiency LED traffic lights for cities that lacked the money, ultimately converting the entire state.

Significantly, California adopted regulations so that utility company profits are not tied to how much electricity they sell. This is called "decoupling." It also allowed utilities to take a share of any energy savings they help consumers and businesses achieve. The bottom line is that California utilities can make money when their customers save money. That puts energy-efficiency investments on the same competitive playing field as generation from new power plants.

The cost of efficiency programs has averaged 2 to 3 cents per avoided kilowatt hour, which is about one-fifth the cost of electricity generated from new nuclear, coal, and natural gas–fired plants. And, of course, energy efficiency does not require new power lines and does not generate greenhouse-gas emissions or long-lived radioactive waste. While California is far more efficient than the rest of the country, the state still thinks that with an even more aggressive effort, it can achieve as much additional electricity savings by 2020 as it has in the last three decades.

Serious energy efficiency is not a one-shot resource, where you pick the low-hanging fruit and you're done. In fact, the fruit grows back. The efficiency resource never gets exhausted because technology keeps improving and knowledge spreads to more people.

The best corporate example is Dow Chemical's Louisiana division, consisting of more than twenty plants. In 1982, the division's energy manager, Ken Nelson, began a yearly contest to identify and fund energy-saving projects. Some of the projects were simple, like more efficient compressors and motors, or better insulation for steam lines. Some involved more sophisticated thermodynamic "pinch" analysis, which allows engineers to figure out where to place heat exchangers to capture heat emitted in one part of a chemical process and transfer it to a different part of the process where heat is needed. His success was nothing short of astonishing.

The first year of the contest had twenty-seven winners requiring a total capital investment of $1.7 million with an average annual return on investment of 173 percent. Many at Dow felt that there couldn't be others with such high returns. The skeptics were wrong. The 1983 contest had thirty-two winners requiring a total capital investment of $2.2 million and a 340 percent return—a savings of $7.5 million in the first year and every year after that. Even as fuel prices declined in the mid-1980s, the savings kept growing. The average return to the 1989 contest was the highest ever, an astounding 470 percent in 1989—a payback of 11 weeks that saved the company $37 million a year.

You might think that, after ten years and nearly 700 projects, the 2,000 Dow employees would be tapped out of ideas. Yet the contest in 1991, 1992, and 1993 each had in excess of 120 winners with an average return on investment of 300 percent. Total savings to Dow from just those projects exceeded $75 million a year.

When I worked at the Department of Energy in the mid-1990s, we hired Nelson, who had recently retired from Dow, to run a "return on investment" contest to reduce DOE's pollution. As they were at Dow, many DOE employees were skeptical such opportunities existed. Yet the first two contest rounds identified and funded eighteen projects that cost $4.6 million and provided the department $10 million in savings every year, while avoiding more than 100 tons of low-level radioactive pollution and other kinds of waste. The DOE's regional operating officers ended up funding 260 projects costing $20 million that have been estimated to achieve annual savings of $90 million a year.

Economic models greatly overestimate the cost of carbon mitigation because economists simply don't believe that the economy has lots of high-return energy-efficiency opportunities. In their theory, the economy is always operating near efficiency. Reality is very different from economic models.

In my five years at DOE, working with companies to develop and deploy efficient and renewable technologies, and then in nearly a decade of consulting with companies in the private sector, I never saw a building or factory that couldn't cut electricity consumption or greenhouse-gas emissions 25 percent to 50 percent with rapid payback (under four years). My 1999 book, *Cool Companies*, detailed some 100 case studies of companies that have done just that and made a great deal of money.

There are many reasons why most companies don't match what the best companies do. Until recently, saving energy has been a low priority for most of them. Most utilities, as noted, have little or no incentive to help customers save energy. Funding for government programs to help companies adopt energy-saving strategies has been cut under the Bush administration.

Government has a very important role in enabling energy savings. The office of Energy Efficiency and Renewable Energy at the U.S. Department of Energy has lots of (underfunded) programs that deliver savings every day. Consider, for instance, Chrysler's St. Louis complex, which recently received a DOE Save Energy Now energy assessment. Using DOE software, Chrysler identified a variety of energy-saving measures and saved the company $627,000 a year in energy costs—for an upfront implementation cost of only $125,000.

The key point for policy makers now is that we have more than two decades of experience with successful state and federal energy-efficiency programs. We know what works. As California energy commissioner Art Rosenfeld—a former DOE colleague and the godfather of energy efficiency—put it in a recent conversation, "A lot of technology and strategies that are tried and true in California are waiting to be adopted by the rest of country."

So how do we overcome barriers and tap our nearly limitless efficiency resource? Obviously, the first thing would be to get all the states to embrace smarter utility regulations, which is a core strategy of Barack Obama's plan to reduce greenhouse gases. But how does the federal government get all the states to embrace efficiency?

We should establish a federal matching program to co-fund state-based efficiency programs, with a special incentive to encourage states without an efficiency program to start one. This was a key recommendation of the End-Use Efficiency Working Group to the Energy Future Coalition, a bipartisan effort to develop consensus policies, in which I participated. The first year should offer $1 billion in federal matching funds, then $2 billion, $3 billion, $4 billion, and finally stabilizing at $5 billion. This will give every state time to change their regulations and establish a learning curve for energy efficiency.

This program would cost $15 billion in the first five years, but save several times that amount in lower energy bills and reduced pollution. Since the next president will put in place a cap-and-trade system for greenhouse gases, the revenues from auctioning the emissions permits can ultimately be used to pay for the program.

We should restore a federal focus on the energy-intensive industries, such as pulp and paper, steel, aluminum, petroleum refining, and chemicals. They account for 80 percent of energy consumed by U.S. manufacturers and 90 percent of the hazardous waste. They represent the best chance for increasing efficiency while cutting pollution. Many are major emitters of greenhouse gases other than carbon dioxide. A 1993 analysis for the DOE found that a 10 to 20 percent reduction in waste by American industry would generate a cumulative increase of $2 trillion in the gross domestic product from 1996 to 2010. By 2010, the improvements would be generating 2 million new jobs.

For these reasons, in the 1990s, the Energy Department began forming partnerships with energy-intensive industries to develop clean technologies. We worked with scientists and engineers to identify areas of joint research into technologies that would simultaneously save energy, reduce pollution, and increase productivity. The Bush administration slashed funding for this program by 50 percent—and tried to shut it down entirely.

Indeed, conservatives in general have cut the funding or shut down entirely almost all federal programs aimed at deploying energy-efficient technologies. Conservatives simply have a blind spot when it comes to energy efficiency and conservation, seeing them as inconsequential "Jimmy Carter programs."

I recently testified at a Senate Environment and Public Works Committee hearing on nuclear power and spoke about how alternative technologies, particularly energy efficiency, were a much better bet for the country. Senator George Voinovich (R-Ohio) said this was "poppycock," and then asked all the pro-nuclear witnesses to address the question, "If nuclear power is so uncompetitive, why are so many utilities building reactors?"

Voinovich apparently has forgotten about the massive subsidies he himself voted to give the nuclear industry in 2005. He seems to be unaware that states like Florida allow utilities to sharply raise electric rates years in advance of a nuclear plant delivering even a single electron to customers. If you could do that same forward-pricing with energy efficiency, we would never need to build another polluting plant.

Although he is a senior member of the Senate and a powerful voice on energy and climate issues, Voinovich ignores the basics of the electricity business; namely, that a great many utilities have a huge profit incentive to build even the most expensive power plants, since they can pass all costs on to consumers while retaining a guaranteed profit. But they have a strong disincentive from investing in much less costly efforts to reduce electricity demand, since that would eat into their profits.

President Obama must challenge the public service commission in every state to allow utilities to receive the same return on energy efficiency as they are allowed to receive on generation. That single step could lead the country the furthest in solving our ever-worsening climate and energy problems.

The Technology That Will Save Humanity: The Solar Energy You Haven't Heard of Is the One Best Suited to Generate Clean Electricity for Generations to Come

April 14, 2008

One of the oldest forms of energy used by humans—sunlight concentrated by mirrors—is poised to make an astonishing comeback. I believe it will be the most important form of carbon-free power in the twenty-first century. That's because it's the only form of clean electricity that can meet all the demanding requirements of this century.

Certainly we will need many different technologies to stop global warming. They include electric cars, plug-in hybrids, wind turbines, and solar photovoltaics, which use sunlight to make electricity from solid-state materials like silicon semiconductors. Yet after speaking with energy experts and seeing countless presentations on all forms of clean power, I believe the one technology closest to being a silver bullet for global warming is the other solar power: solar thermal electric, which concentrates the sun's rays to heat a fluid that drives an electric generator. It is the best source of clean energy to replace coal and sustain economic development. I bet that it will deliver more power every year this century than coal with carbon capture and storage—for much less money and with far less environmental damage.

Clearly, the world needs a massive amount of carbon-free electricity by 2050 to stabilize greenhouse gas emissions. The industrialized countries need to cut their carbon dioxide emissions from electricity generation by more than 80 percent in four decades. Developing countries need to find a way to raise living standards without increasing electricity emissions in the short term, and then reduce those emissions sharply. And, over the next few decades, the world needs to switch to a ground transportation system whose primary fuel is clean electricity.

This electricity must meet a number of important criteria. It must be affordable: New electricity generation should cost at most about 10 cents per kilowatt hour, a price that would probably beat nuclear power and would certainly beat coal with carbon capture and storage, if the latter even proves practical on a large scale. The electricity cannot be intermittent and hard to store, as is energy from wind power and solar photovoltaics. We need power that either stays constant day and night or, even better, matches electricity demand, which typically rises in the morning, peaks in the late afternoon, and lasts late into the evening.

This carbon-free electricity must provide thousands of gigawatts of power and make use of a low-cost fuel that has huge reserves accessible to both industrialized and developing countries. It should not make use of much freshwater or arable land, which are likely to be scarce in a climate-changed world with 3 billion more people.

Solar electric thermal, also known as concentrated solar power (CSP), meets all these criteria. A technology that has the beauty of simplicity, it has proved effective for generations. As the website of CSP company Ausra illustrates, solar thermal has a long and fascinating history.

Back around 700 B.C., the Chinese first used "burning mirrors" to ignite firewood. In 230 B.C., a colleague of Archimedes built a parabolic mirror, which focuses the sun's rays to a single point, also better for starting fires. Around 212 B.C., Archimedes supposedly had Greek soldiers use their bronze shields to concentrate the sunlight on Roman ships and set them on fire.

In the fifteenth century, the Italians used burning mirrors to solder copper sections of the Santa Maria del Fiore cathedral. Leonardo da Vinci's notebooks contain many designs for solar concentrators, including some for industrial purposes, because he worried about the destruction of the Earth's vast forests in humanity's search for fuel.

In the 1860s and 1870s, Augustin Mouchot built the first dish-shaped reflector that ran a heat engine, and he used solar thermal to heat a boiler that ran an ice maker. His assistant demonstrated a printing press running on concentrated solar. But all this work came to naught because of the general lack of direct sunlight in France and the abundance of cheap coal, which became a primary energy source for the Industrial Revolution.

A Swedish immigrant to America, John Ericsson, developed a motor driven by parabolic trough mirrors in 1870. In 1909, H. E. Wilsie added a critical component: a system for storing solar energy for when the sun did not shine. Heat is much easier to store than electricity, a fact that gives CSP a crucial—maybe *the* crucial—advantage over wind and solar photovoltaics.

In 1913, an American, Frank Shuman, installed a 55-kilowatt CSP water-pumping station using parabolic mirrors in Meadi, Egypt. The mirrors focused the sun on tubes whose heated fluid ran an engine to make electricity. This was perhaps the first commercial CSP plant. But it was shut down at the start of WWI, and, as Ausra notes, "the plant was never restarted because of the discovery of cheap oil in the Middle East."

In the 1960s, the Italians developed two of the key CSP designs used today. The first uses a linear mirror to focus the light on a long tube, allowing the mirrors to be flat, cheaper to build, and less exposed to the wind. In the second, called a power tower, many mirrors move in two dimensions, focusing on a central tower that holds the engine.

The 1970s oil shocks led to the first commercial developer of U.S. solar thermal electric projects, Luz International. The company built and sold nine solar plants in California's Mojave Desert. The plants circulated oil in pipes, heating it to 700 degrees with long parabolic mirrors; the oil boiled water to drive a steam turbine. Although the technology functioned well, Luz was forced to file for bankruptcy in 1991. The reasons included uncertainty in the market, a delay of federal and state tax breaks, and the lack of economic value derived from environmental benefits.

For more than a decade, those barriers, coupled with low natural-gas prices, kept CSP moribund. The technology got a huge boost in 2004, when Spain approved a guaranteed price, a "feed-in tariff," for CSP. That led to an explosion of Spanish CSP, starting with a power tower near Seville, and a plant outside Granada, the first parabolic trough system in Europe, which should be running later this year.

In this country, soaring gas prices and renewable portfolio standards have sparked a resurgence. In 2006, the Arizona Public Service Co. dedicated the first new CSP plant in the United States in two decades—a 1-megawatt concentrated solar trough system with an engine used for decades by the geothermal industry. In June 2007, Nevada Solar One, the state's first CSP plant, went online. On 275 acres near Boulder City, it provides 64 MW of electricity from 98 percent solar power and 2 percent natural gas. And in California, PG&E has created deals with three major CSP companies to generate electricity for the Golden State. Another ten plants are in the advanced planning stages in the Southwest, along with nine plants in countries that include Israel, Mexico, and China.

The key attribute of CSP is that it generates primary energy in the form of heat, which can be stored 20 to 100 times more cheaply than electricity—and with far greater efficiency. Commercial projects have already demonstrated that CSP systems can store energy by heating oil or molten salt, which can retain the heat for hours. Ausra and other companies are working on storing the heat directly with water in the tubes, which would significantly lower cost and avoid the need for heat exchangers.

CSP costs have already begun to decline as production increases. According to a 2008 Sandia National Laboratory presentation, costs are projected to drop to 8 to 10 cents per kilowatt hour when capacity exceeds 3,000 MW. The world will probably have double that capacity by 2013. The price drop will likely occur even if the current high prices for raw materials like steel and concrete continue (prices that also affect the competition, like wind, coal, and nuclear power).

Since all three remaining presidential candidates endorse a cap on carbon dioxide emissions coupled with a system for trading emissions permits, carbon dioxide will likely have a significant price within a few years. And that means the economics of carbon-free CSP will only get better. Improvements in manufacturing and design, along with the possibility of higher temperature operation, could easily bring the price down to 6 to 8 cents per kilowatt hour.

CSP makes use of the most abundant and free fuel there is, sunlight, and key countries have a vast resource. Solar thermal plants covering the equivalent of a 92-by-92-mile square grid in the Southwest could generate electricity for the entire United States. Mexico has an equally enormous solar resource. China, India, southern Europe, North Africa, the Middle East, and Australia also have huge resources.

CSP plants can also operate with a very small annual water requirement because they can be air-cooled. And CSP has some unique climate-friendly features. It can be used effectively for desalinating brackish water or seawater. That is useful for many developing countries today, and it's a must-have for tens if not hundreds of millions of people if we don't act in time to stop global warming and dry out much of the planet. Such desertification would, ironically, mean even more land ideal for CSP.

The technology has no obvious bottlenecks and uses mostly commodity materials—steel, concrete, and glass. The central component, a standard power system routinely used by the natural gas industry today, would create steam to turn a standard electric generator. Plants can be built rapidly—in two to three years—much faster than nuclear plants. It would be straightforward to build CSP systems at whatever rate industry and governments needed, ultimately 50 to 100 gigawatts a year growth or more.

So what do we need to do to ramp up CSP? Interestingly, most CSP executives don't talk much about the need for government R&D. They mostly need policies aimed at creating initial market demand that would help bring down costs quickly over the next several years. One such policy is a so-called national renewable portfolio standard, which would require utilities to get a minimum percentage of their electricity from new renewable forms of power, or purchase such power from other utilities. After that, the typical manufacturing learning curves and economies of scale—plus a market price for carbon dioxide set by the cap-and-trade system—should do the rest.

That means Congress and the president must renew the 30 percent solar energy investment tax credit through 2016. After all, it's the least they can do. From 2002 to 2007, fossil fuels received almost $14 billion in electricity-related tax subsides, whereas renewables received less than $3 billion. From 1948 to today, nuclear energy R&D exceeded $70 billion, whereas R&D for renewables was about $10 billion.

The United States has already lost the leadership it had in solar photovoltaics and wind, thanks to deep budget cuts by President Reagan and the Newt Gingrich–led Congress. By 2010, China will be the top manufacturer of photovoltaic cells and wind turbines. Must we also abandon our historical leadership in CSP to conservative doctrine? Other countries—particularly Spain but also Israel and Australia—are poised to be dominant. And China, which has already begun importing coal and pursuing CSP projects, will not be far behind. CSP could well be one of the major job-creating industries of the century.

Every other major country aggressively supports clean tech industries with subsidies and mandates. But our Congress and president can't even agree on a requirement for 10 percent of U.S. energy to be from renewable sources—far less than most European countries and half our own states. We should have a federal standard requiring U.S. utilities to get 20 percent of their power from renewables by 2020.

Another useful incentive would be loan guarantees, a program that could be retired once we have a price for carbon dioxide. CSP has no fuel cost, and low operations and maintenance costs, but it has high upfront capital costs. Loan guarantees can reduce the risks of the first big plants at little or no cost to the taxpayer. The United States should also insist that CSP be a high priority for development projects by the Global Environmental Facility and the World Bank.

Finally, we will need more electric transmission in this country. The good news is that because it matches the load most of the day and has cheap storage, CSP can share power lines with wind farms. When the country gets serious about global warming, we will need to get serious about building a transmission system for a low-carbon economy.

If we are smart, the United States can be the economic leader here. We can accelerate the deployment of a technology that may be critical to saving humanity from a ruined climate.

The Breakthrough Technology Illusion

April 6, 2009

This post will explain why some sort of massive government Apollo program or Manhattan project to develop new breakthrough technologies is *not* a priority component of the effort to stabilize at 350 to 450 ppm.

Put more quantitatively, the question is: What are the chances that by 2050 multiple (4 to 8+) carbon-free technologies that do not exist today can *each* deliver the equivalent of 350 gigawatts baseload power (~2.8 billion megawatt-hours a year) and/or 160 billion gallons of gasoline *cost-effectively*?‡ For the record, the United States consumed about 3.7 billion MW-hrs in 2005 and about 140 billion gallons of motor gasoline.

Put that way, the answer to the question is painfully obvious: "two chances—slim and none." Indeed, I have repeatedly challenged readers and listeners over the years to name even a single technology breakthrough with such an impact in the past three decades, after the huge surge in energy funding that followed the energy shocks of the 1970s. Nobody has ever named a single one that has even come close.

Yet somehow the government is not just going to invent one TILT (Terrific Imaginary Low-carbon Technology) in the next few years, we are going to invent several TILTs. Seriously. Hot fusion? No. Cold fusion? As if. Space solar power? Come on, how could that ever compete with concentrated solar power? Hydrogen? It ain't even an energy source, and after billions of dollars of public and private research in the last fifteen years—including several years running of being the single biggest focus of the DOE office on climate solutions I once ran—it still has actually no

‡ That is about half of a stabilization wedge.

chance whatsoever of delivering a major cost-effective climate solution by midcentury if ever.

I don't know why the breakthrough crowd can't see the obvious—so I will elaborate here. I will also discuss a major study that explains why deployment programs are so much more important than R&D at this point. Let's keep this simple:

- To stabilize below 450 ppm, we need to deploy by 2050 some twelve to fourteen stabilization wedges (each delivering 1 billion tons a year of avoided carbon) covering both efficient energy use and carbon-free supply. The technologies we have today, plus a few that are in the verge of being commercialized, can provide the needed low-carbon energy.
- Myriad energy-efficient technologies are already cost-effective today. Breaking down the barriers to their deployment now is much, much more important than developing new "breakthrough" efficient TILTs, since those would simply fail in the marketplace because of the same barriers. Cogeneration (combined heat and power) is perhaps the clearest example of this.
- On the supply side, deployment programs (coupled with a price for carbon) will always be much, much more important than R&D programs because new technologies take an incredibly long time to achieve mass-market commercial success. New supply TILTs would not simply emerge at a low cost. They need volume, volume, volume—steady and large increases in demand over time to bring the cost down, as I discuss at length later in this section.
- No existing *or* breakthrough technology is going to beat the price of power from a coal plant that has already been built—the only way to deal with those plants is a high price for carbon or a mandate to shut them down. Indeed, that's why we must act immediately not to build those plants in the first place.
- If a new supply technology can't deliver half a wedge, it won't be a big player in achieving 350 to 450 ppm.

For better or worse, we are stuck through 2050 with the technologies that are commercial today (like solar thermal electric) or that are very nearly commercial (like plug-in hybrids).

I have discussed most of this at length in previous posts, so I won't repeat all the arguments here. Let me just focus on a few key points. A critical historical fact was explained by Royal Dutch/Shell, in their 2001 scenarios for how energy use is likely to evolve over the next five decades (even with a carbon constraint): "Typically it has taken twenty-five years after commercial introduction for a primary energy form to obtain a 1 percent share of the global market."

Note that this tiny toehold comes twenty-five years after *commercial* introduction. The first transition from scientific breakthrough to commercial introduction may itself take decades. We still haven't seen commercial introduction of a hydrogen fuel cell car and have barely seen any commercial fuel cells—170 years after they were first invented.

This tells you two important things. First, new breakthrough energy technologies simply don't enter the market fast enough to have a big impact in the time frame we care about. We are trying to get 5 percent to 10 percent shares—or more—of the global market for energy, which means massive deployment by 2050 (if not sooner).

Second, if you are in the kind of hurry we are all in, then you are going to have to take unusual measures to deploy technologies far more aggressively than has ever occurred historically. That is, speeding up the deployment side is much more important than generating new technologies. Why? Virtually every supply technology in history has a steadily declining cost curve, whereby greater volume leads to lower cost in a predictable fashion because of economies of scale and the manufacturing learning curve.

Why Deployment Now Completely Trumps Research

A major 2000 report by the International Energy Agency (IEA), *Experience Curves for Energy Technology Policy*, has a whole bunch of experience curves for various energy technologies. Let me quote some key passages:

> Wind power is an example of a technology which relies on technical components that have reached maturity in other technological fields. . . . Experience curves for the total process of producing electricity from

> wind are considerably steeper than for wind turbines. Such experience curves reflect the learning in choosing sites for wind power, tailoring the turbines to the site, maintenance, power management, etc., which all are new activities.

Or consider PV:

> Existing data show that experience curves provide a rational and systematic methodology to describe the historical development and performance of technologies. . . .
>
> The experience curve shows the investment necessary to make a technology, such as PV, competitive, but it does not forecast when the technology will break even. The time of break even depends on deployment rates, which the decision maker can influence through policy. With historical annual growth rates of 15%, photovoltaic modules will reach break-even point around the year 2025. Doubling the rate of growth will move the break-even point 10 years ahead to 2015.
>
> Investments will be needed for the ride down the experience curve, that is for the learning efforts which will bring prices to the break-even point. An indicator for the resources required for learning is the difference between actual price and break-even price, i.e., the additional costs for the technology compared with the cost of the same service from technologies which the market presently considers cost-efficient. We will refer to these additional costs as learning investments, which means that they are investments in learning to make the technology cost-efficient, after which they will be recovered as the technology continues to improve.

Here is a key conclusion:

> For major technologies such as photovoltaics, wind power, biomass, or heat pumps, *resources provided through the market dominate the learning investments.* Government deployment programmes may still be needed to stimulate these investments. The government expenditures for these programmes will be included in the learning investments.

Obviously government R&D, and especially first-of-a-kind demonstration programs, are critical before the technology can be introduced to the marketplace on a large scale—and I'm glad Obama had doubled spending in this area. But, we "expect learning investments to become the dominant resource for later stages in technology development, where the objectives are to overcome cost barriers and make the technology commercial."

We are really in a race to get technologies into the learning curve phase: "The experience effect leads to a competition between technologies to take advantage of opportunities for learning provided by the market. To exploit the opportunity, the emerging and still too expensive technology also has to compete for learning investments."

In short, you need to get from first demonstration to commercial introduction as quickly as possible to be able to then take advantage of the learning curve before your competition does. Again, that's why if you want mass deployment of the technology by 2050, we are mostly stuck with what we have today or very soon will have. Some breakthrough TILT in the year 2025 will find it exceedingly difficult to compete with technologies like CSP or wind that have had decades of such learning.

And that is why the analogy of a massive government Apollo program or Manhattan project is so flawed. Those programs were to create unique noncommercial products for a specialized customer with an unlimited budget. Throwing money at the problem was an obvious approach. To save a livable climate we need to create mass-market commercial products for lots of different customers who have limited budgets. That requires a completely different strategy.

Finally, it should be obvious, but it apparently isn't, so I'll quote the IEA again:

> The risk of climate change, however, poses an externality which might be very substantial and costly to internalise through price alone. Intervening in the market to support a climate-friendly technology that may otherwise risk lock-out may be a legitimate way for the policy maker to manage the externality; the experience effect thus expands his policy options. For example, carbon taxes in different sectors of the economy can activate the learning for climate-friendly technologies by raising the break-even price.

So, yes, a price for carbon is exceedingly important—more important, as I have argued, than funding the search for TILTs.

The Breakthrough Bunch

The *New York Times*'s Andy Revkin interviewed a whole bunch of people who think we need "massive public investments" and breakthroughs. Revkin writes: "Most of these experts also say existing energy alternatives and improvements in energy efficiency are simply not enough."

The devil is always in the details of the quotes—especially since everybody I know wants more federal investments on low-carbon technologies. And, of course, some of the folks Revkin quotes are long-time delayers, like W. David Montgomery of Charles River Associates—who has testified many times that taking strong action on climate change would harm the economy. He says stabilizing temperatures by the end of the century "will be an economic impossibility without a major R&D investment." Well, of course he would. In any case, we don't have until the end of the century—yes, it would certainly be useful to have new technologies in the second half of this century, but the next couple of decades are really going to determine our fate.

Revkin quotes my friend Jae Edmonds as saying we need to find "energy technologies that don't have a name yet." Jae and I have long disagreed on this, and he is wrong. His economic models have tended to assume a few major breakthroughs in a few decades and that's how he solves the climate problem. Again, I see no evidence that that is a plausible solution nor that we have the time to wait and see.

I would estimate that the actual federal budget today that goes toward R&D breakthroughs that could plausibly deliver a half wedge or more by 2050 (i.e., not fusion, not hydrogen) is probably a few hundred million dollars at most. I wouldn't mind raising that to a billion dollars a year. But I wouldn't spend more, especially as long as the money was controlled by a Congress with its counterproductive earmarks. I could probably usefully spend 10 times that on deployment (not counting tax policy), again as long as the money was not controlled by Congress. Since that may be difficult if not impossible to arrange, we have to think hard about what the size of a new federal program might be.

Yet another reason we don't need a sort of massive government Apollo program or Manhattan project is that the venture-capital community has massively ramped up cleantech spending just where it is most needed—key low-carbon technologies that have a serious chance of becoming commercial in the next three to five years.

Some critics have said that my proposed fourteen wedges require betting the future on "some fantastically delusional expectations of the possibilities of policy implementation" and that my allegedly "fuzzy math explains exactly why innovation must be at the core of any approach to mitigation that has a chance of succeeding." Well, we've seen my math wasn't fuzzy.

But you tell me, what is more delusional—(1) that we take a bunch of commercial or very near commercial technologies and rapidly accelerate their deployment to wedge-scale over the next four decades or (2) that in the same exact time frame, we invent a bunch of completely new technologies "that don't have a name yet," commercialize them, and then rapidly accelerate them into the marketplace so they achieve wedge scale?

And so I assert again, the vast majority—if not all—of the wedge-sized solutions for 2050 will come from technologies that are now commercial or very soon will be. And federal policy must be designed with that understanding in mind. So it seems appropriate to end this post with an excerpt from the Conclusion of the IEA report:

> A general message to policy makers comes from the basic philosophy of the experience curve. Learning requires continuous action, and future opportunities are therefore strongly coupled to present activities. If we want cost-efficient, CO_2-mitigation technologies available during the first decades of the new century, these technologies must be given the opportunity to learn in the current marketplace. Deferring decisions on deployment will risk lock-out of these technologies, i.e., lack of opportunities to learn will foreclose these options making them unavailable to the energy system . . .
>
> The low-cost path to CO_2-stabilisation requires large investments in technology learning over the next decades. The learning investments are provided through market deployment of technologies not yet commercial, in order to reduce the cost of these technologies and make them competitive with conventional fossil-fuel technologies.

> Governments can use several policy instruments to ensure that market actors make the large-scale learning investments in environment-friendly technologies. Measures to encourage niche markets for new technologies are one of the most efficient ways for governments to provide learning opportunities. The learning investments are recovered as the new technologies mature, illustrating the long-range financing component of cost-efficient policies to reduce CO_2 emissions. The time horizon for learning stretches over several decades, which require long-term, stable policies for energy technology.

Deployment, deployment, deployment, R&D, deployment, deployment, deployment.

Introduction to Climate Economics: Why Even Strong Climate Action Has Such a Low Total Cost— One Tenth of a Penny on the Dollar

March 30, 2009

In its definitive 2007 synthesis report of the scientific literature, the Intergovernmental Panel on Climate Change (IPCC) concluded:

> In 2050, global average macro-economic costs for mitigation towards stabilisation between 710 and 445ppm CO_2-eq are between a 1% gain and 5.5% decrease of global GDP. This corresponds to slowing average annual global GDP growth by *less than 0.12 percentage points.*

So global GDP drops by less than 0.12 percent per year—about one tenth of a penny on the dollar—even in the 445 ppm CO_2-eq case (through 2050). And this is for stabilization at 445 ppm CO_2-eq (CO_2 equivalent—all greenhouse gases), which is stabilization at 350 ppm CO_2.

And 350 ppm has a very good chance of averting the incalculable cost of catastrophic global warming impacts to the next fifty generations, which means the cost of action is far, far less than the cost of inaction.

The IPCC's conclusion—and every single word in the report—was signed off on by more than 100 nations including China as well as the Bush Administration. Nor is this an especially controversial conclusion, at least among the few groups that have done comprehensive global economic and energy modeling:

- McKinsey concluded in 2008 that stabilizing at 450 ppm has a net cost near zero. (See cost curve on the next page where the savings from the measures on the left side of the curve nearly pay the cost of the measures on the right side.)
- The International Energy Agency concluded in 2009 that total cost of stabilization is low whereas failure to act could lead to warming of 10°F.

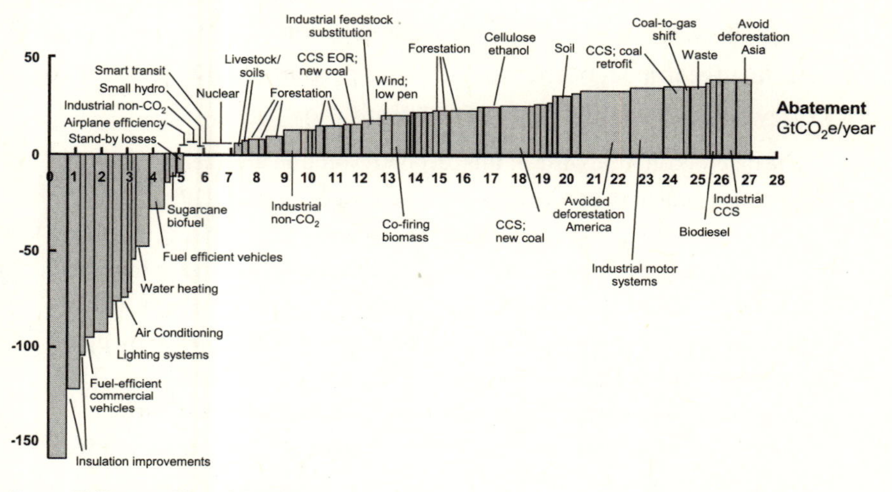

Source: McKinsey and Vattenfall analysis

How can the world's leading governments *and* scientific experts *and* McKinsey *and* the traditionally conservative International Energy Agency agree that we can avoid catastrophe for such a small cost?

Because that's what the scientific and economic literature—and real-world experience—says. The IPCC summary report, which is, after all, primarily a literature review, notes:

> Both bottom-up and top-down studies indicate that there is high agreement and much evidence of substantial economic potential for the mitigation of global GHG emissions over the coming decades that could offset the projected growth of global emissions or reduce emissions below current levels.

In fact, the bottom-up studies—the ones that look technology by technology, which I believe are more credible—have even better news:

> Bottom-up studies suggest that mitigation opportunities with net negative costs have the potential to reduce emissions by around 6 $GtCO_2$-eq/yr in 2030.

Wow! A 20 percent reduction in global emissions might be possible over a quarter century with net economic benefits!

The technology-by-technology cost-curve from McKinsey demonstrates this finding more concretely. Whereas the IPCC merely says that 450 ppm could be achieved for a total GDP reduction of less than 3 percent in 2030 (the cumulative impact of the less than 0.12 percent of GDP per year cost), McKinsey believes it could be even less costly:

> The macroeconomic costs of this carbon revolution are likely to be manageable, being in the order of 0.6–1.4 percent of global GDP by 2030. To put this figure in perspective, if one were to view this spending as a form of insurance against potential damage due to climate change, it might be relevant to compare it to global spending on insurance, which was 3.3 percent of GDP in 2005. Borrowing could potentially finance many of the costs, thereby effectively limiting the impact on near-term GDP growth. In fact, depending on how new low-carbon infrastructure is financed, the transition to a low-carbon economy may increase annual GDP growth in many countries.

I want to be clear here that stabilizing at 445 ppm CO_2-equivalent does require a significant annual investment, as the IEA analysis shows. The IEA puts the investment at $45 trillion, which sounds like an unimaginably large amount of money—but spread over more than four decades and compared to the world's total wealth during that time, it is literally a drop in the bucket—1.1 percent or 1 part in 90 of the world's total wealth.

Indeed, the IEA notes that one reason the dollar value of the investment is so high is "in part due to the declining value of the dollar."[§]

And while the additional investments seem high, "they do not represent net costs." They are not a pure negative hit to global GDP. That's because "technology investments in energy efficiency" and many low-carbon power sources "reduce fuel requirements." In all the scenarios the IEA considers:

> The estimated total undiscounted fuel cost savings for coal, oil, and gas over the period to 2050 are greater than the additional investment required (valuing these fuels at Baseline prices). If we discount at 3%, fuel savings exceed additional investment needs in the ACT Map scenario [in which CO_2 emissions in 2050 only return to 2005 levels].

---

§ Note to self: How diabolical of President Bush—by weakening our economy he increased the total dollar cost of action on climate, thus encouraging inaction!

But don't we need new technologies? Of course, but we don't need—and can't afford—to sit on our hands when we now have so many cost-effective existing technologies. The IPCC finds:

> There is *high agreement* and *much evidence* that all stabilisation levels assessed can be achieved by deployment of a portfolio of technologies that are either currently available or expected to be commercialised in coming decades, assuming appropriate and effective incentives are in place for their development, acquisition, deployment, and diffusion and addressing related barriers.

Yes, we need to do two things at once: aggressively deploy existing technology (with carbon prices and government standards) and aggressively finish developing and commercializing key technologies and systems that are in the pipeline. Anyone who argues for just doing the latter is disputing a very broad consensus—and is neither pragmatic nor centrist.

McKinsey finds 70 percent of the total 2030 emissions reduction potential (below $60 a ton of CO_2 equivalent) is "not dependent on new technology." The report notes that "we have been fairly conservative in our assumptions about technological progress in these projections." For instance, the analysis largely ignores the potential of concentrated solar thermal electricity, which is a bit player for their analysis but which will probably be the single biggest supply side low carbon source in reality (as we've seen).

The IEA report does suggest we need major technology advances—but that is mostly for cost reduction in the transportation sector if the price of oil stays frozen at $65 a barrel, which even the IEA doesn't believe any more (see chapter 4).

So the bottom line is that the economic cost of action is low, whereas the cost of inaction is incalculably greater—what exactly is the "price" of 5 feet of sea level rise in 2100, rising 6 or more inches a decade for centuries thereafter, or the price of turning one third of the habited land mass into a dust bowl and losing all of the inland glaciers that provide a significant fraction of water to a billion people? Or the price of losing half the world's species?

And this is without even adding in the various ancillary benefits such as reduced air pollution and avoidance of the huge economic dislocations that are inevitable from peak.

CHAPTER 4

Peak Oil? Consider It Solved

ONCE A FRINGE ISSUE, A PEAK IN production of global oil production is increasingly seen as a near-term reality. Just as scientists are increasingly blunt about the dangers posed of global warming, many energy experts are becoming increasingly blunt about the oil situation.

Dr. Fatih Birol, the chief economist at the once staid and conservative International Energy Agency (IEA), is not soft-pedaling the grim reality to anyone who will listen. As *Science* reported in "World Oil Crunch Looming?":

> We have found that if we want to stand still—that is, continue producing 85 million barrels per day—for the next 22 years, we need new production of 45 million barrels per day to compensate for the decline. That means four Saudi Arabias.

Add on a demand increase of the sort seen the past couple of decades—equivalent to another two Saudi Arabias—and the world will have to work that much harder to meet rising demand, Birol says.

Those six Saudi Arabias do not exist underground—they can only be found in the nation's (and the world's) cars, trucks, buildings, factories, power plants, and farms. America is the Saudi Arabia of wasted energy. And we now know what the winning low-carbon alternative fuel is—electricity.

It Won't Be Easy, but We Can Fix Our Oil and Climate Problems at the Same Time

March 28, 2008

For more than a decade, a fierce debate about peak oil has been raging between those who think a peak in global oil production is at hand and those who think the world is not close to running out of oil. The debate is moot for two reasons. First, the growing threat of global warming requires deep reductions in national and global oil consumption starting now, peak or no peak. Second, relying on unconventional oil like tar sands and liquid coal to make up a supply shortage, as the oilmen say we must, would be climate catastrophe. More supply is not the answer to either our oil or our climate problem—reducing consumption of oil is. And right now we have two feasible solutions: greatly increase our vehicle fuel economy and find alternative fuel sources that are abundant, low-carbon, and affordable.

Make no mistake about it: Soaring global oil consumption has brought the nation and the world to a point of reckoning. Last year, consumption was 86 million barrels a day, up from 78 million in 2002, roughly a 2 percent annual rise. Where is all the demand coming from? Hint: It's not just the rapidly developing countries. From 1995 to 2004, China's annual imports grew by 2.8 million barrels a day. Ours grew 3.9 million. China now sucks up about 6 percent of all global oil exports. We demand 25 percent. American's trade deficit in oil alone is nearing $500 billion a year.

That said, if by 2050, the per capita energy consumption of China and India were to approach that of South Korea, and if the Chinese and Indian populations increase at currently projected rates, those two super giant countries by themselves would consume more oil than the entire world used last year.

This massive, unsustainable consumption has more than peak oil doomsayers like James Kunstler worried. In January, Jeroen van der Veer, chief executive officer of Royal Dutch/Shell, e-mailed his staff that the world will peak in conventional oil and gas within the decade. He wrote: "Shell estimates that after 2015 supplies of easy-to-access oil and gas will no longer keep up with demand." It used to be unheard of for oil executives to talk about limits to oil production. Now it happens all the time.

John Hess, chairman of Hess Corp., a global oil and mineral exploration company, said recently, "An oil crisis is coming in the next ten years. It's not a matter of demand. It's not a matter of supplies. It's both." In October, Christophe de Margerie, CEO of French oil company Total S.A., said that production of even 100 million barrels a day by 2030 will be "difficult." In November, James Mulva, CEO of ConocoPhillips, the third biggest U.S. oil company, told a Wall Street conference: "I don't think we are going to see the supply going over 100 million barrels a day . . . Where is all that going to come from?"

The problem is graver than it appears for one simple reason: Replacing oil in the transportation sector requires strong government action two decades before a peak because of the time needed to replace vehicles and fuel infrastructure. That was the conclusion of a major study funded by the Department of Energy in 2005—yes, the Bush DOE—on "Peaking of World Oil Production." The report notes:

> The world has never faced a problem like this. Without massive mitigation more than a decade before the fact, the problem will be pervasive and will not be temporary. Previous energy transitions (wood to coal and coal to oil) were gradual and evolutionary; oil peaking will be abrupt and revolutionary.

Ouch! The same central point is true about global warming. If we want global carbon dioxide emissions to peak and start declining, the planet will need to start aggressive mitigation policies two decades in advance. We're at about 30 billion tons of annual CO_2 emissions and rising 3 percent per year. By 2020, we'll be over 40 billion tons annually. If we average more than 18 billion tons of CO_2 a year this century, we risk widespread desertification, sea level rise (of 80 feet or more), and the loss of up to 70 percent of all species.

To preserve the livability of the planet, we must cut liquid fossil fuel use more than 50 percent by 2050. That is a central reason that more supply is not the solution to peak oil. That is why it is crucial we don't adopt the strategy that most in the oil industry prefer for dealing with the peak in conventional oil—ramping up unconventional oil. Most of the major forms of unconventional oil will make global warming worse—and some would make a climate catastrophe inevitable.

The world has a number of viscous oils called bitumen, heavy oil, and tar sands (or oil sands). There is more recoverable oil in Canada's tar sands than there is conventional oil in Saudi Arabia. Tar sands are pretty much the heavy gunk they sound like, and making liquid fuels from them requires huge amounts of energy for steam injection and refining. Canada is currently producing about 1 million barrels of oil a day from the tar sands, and that is projected to triple over the next two decades.

Tar sands are doubly dirty. On the one hand, the energy-intensive conversion of tar sands generates two to four times the amount of greenhouse gases per barrel of final product as the production of conventional oil. On the other hand, Canada's increasing use of natural gas to exploit the tar sands is one reason that its exports of natural gas to the United States are projected to shrink in the coming years. So instead of selling clean-burning natural gas to the United States, which we could use to stop the growth of carbon-intensive coal generation, Canada will provide us with a more carbon-intensive oil to burn in our cars. That's lose-lose.

Even more oil can probably be recovered from shale, a claylike rock, than from the tar sands. Most of the world's shale is found in the United States, notably in Colorado and Utah. After the oil shocks of the 1970s, billions were spent exploring the possibility of shale oil, but those efforts were abandoned in the 1980s when oil prices collapsed. Shale does not contain much energy per pound: It has one-tenth the energy of crude oil and one-fourth that of recycled phone books. Converting shale to oil requires a huge amount of energy—possibly as much as 1,200 megawatts of generating capacity to produce 100,000 barrels per day. What a waste of energy just to create a fuel that would spew more greenhouse gases into the air when burned in a car. We must leave the shale in the ground.

The recovery of conventional oil from a well can be enhanced by injecting carbon dioxide into the reservoir. Estimates for potential recovery are 300 to 600 billion barrels. When carbon capture and storage from coal generation becomes commonplace—which might occur as soon as two decades from now—we may be awash in carbon dioxide that could be diverted to enhanced oil recovery. It would be a tragedy if that carbon dioxide was not put into deep underground aquifers (permanently reducing the amount of heat-trapping gas in the atmosphere), but instead used to extract more fossil fuels from the ground (which would ultimately release carbon dioxide into the atmosphere when burned in internal combustion engines). Again, more oil supply doesn't solve the climate problem.

Coal can be converted to diesel fuel using a chemical conversion process called Fischer-Tropsch. During World War II, coal gasification and liquefaction produced more than half of the liquid fuel used by the German military. But the process is incredibly expensive.

You need to spend $5 billion just to build a plant capable of producing 80,000 barrels of oil a day (the United States currently consumes more than 21 million barrels a day). You need about five gallons of water for every gallon of diesel fuel that's produced—not a particularly good long-term strategy in a world facing mega-droughts and chronic water shortages. Worse, the total carbon dioxide emissions from coal-to-diesel are about double that of conventional diesel.

You can capture the carbon dioxide from the process and try to store it underground permanently. But that will make an expensive process even more expensive, so it seems unlikely for the foreseeable future, certainly not until carbon dioxide is regulated and has a high price and we have a number of certified underground geologic repositories. More important, even if you capture the CO_2 from the Fischer-Tropsch process, you are still left with diesel fuel, a carbon-intensive liquid that will release CO_2 into the atmosphere once it is burned in an internal combustion engine.

Coal to diesel is a bad idea for the planet. If the United States or China pursues it aggressively, catastrophic climate change will be all but unavoidable.

A 2006 study by the University of California at Berkeley found that meeting the future shortfall of conventional oil with unconventional oil could increase annual emissions by more than 7 billion tons of carbon dioxide for several decades. That would be fatal to any effort to keep average annual emissions this century below 18 billion tons of CO_2. Indeed, it would probably drive us past dangerous tipping points toward CO_2 levels whose consequences have barely been imagined.

Thus we come to one of the biggest questions of our time: Is humanity wise enough not to pursue carbon-intensive alternative fuels, even though pretty much all of them are economically profitable at current oil prices? Let me assume, optimistically, that we are. Let me also assume that we have more than a decade before the peak in conventional oil. We must act now. And by now I mean when we have a new president who actually cares about these issues and believes in government-led solutions to prevent economic losses from a major oil shock and devastating climate change, each of which would cost the United States trillions of dollars.

Clearly we now have only two realistic strategies: increase our vehicle fuel economy and develop affordable alternative fuel sources that are low in carbon. In 2050, the planet may well have 2 billion cars on the road or more, three times the current number. To avoid dramatic climate impacts, we must use at least 60 percent less total liquid fossil fuels—and that assumes we have essentially eliminated carbon dioxide emissions in the electric sector. The average car on the road will need to put out under one-fifth the emissions of current cars, or the equivalent of five times the "miles per gallon" of today.

If we achieve just half of that emissions cut through greater fuel efficiency (and the other half through a low-carbon alternative fuel), we'll need new cars and SUVs in 2040 to get at least 60 miles per gallon. Of course, that assumes people don't drive greater distances, even though they will be wealthier. A nation's per capita wealth has historically correlated with vehicle miles traveled.

Increased fuel economy can be achieved either by mandates, such as the corporate average fuel economy (CAFE) standards, or by higher prices achieved through higher taxes. Certainly, new technology can help. But no country has ever substantially increased its fuel economy with new technology without relying on much tougher fuel economy standards, higher prices or both. I have never been a big fan of higher gasoline taxes, not just because they are a political nonstarter, but also because you would have to jack up taxes an absurd amount to get the desired impact. So that leaves tougher standards—mandates. Even Europe, with much higher gasoline taxes than us, uses mandates.

Late last year, after some two decades of trying, Congress passed a new 35-mpg standard—after tough fighting with conservatives in both parties. That great achievement will take us in 2020 to where the Chinese are now (but not to where Japan and Europe were six years ago). It is worth noting that China has a minimum-allowable efficiency standard, not a "fleet-average" standard like ours. As the *Toronto Star* nicely explained: "No gasoline-powered car assembled in North America would meet China's current fuel-efficiency standard."

Since we're being optimistic, let's assume we can get fuel economy standards for cars and SUVs of 60 miles per gallon by 2030. We would still need more than half of vehicle fuel to be zero carbon. And for that only one alternative fuel is even remotely plausible—carbon-free electricity. Plug-in hybrids and electric cars are the cars of the future, especially as a climate solution. What's more, with plug-ins and electric cars on the roads, oil peakers like Kunstler—who has claimed that when the oil runs dry, suburbia "will become untenable" and "we will have to say farewell to easy motoring"—can relax.

Suppose that by 2020, oil blew past $300 a barrel and gasoline rose to $9 dollars a gallon (still not much higher than current gasoline prices in England). You could replace your car with a plug-in hybrid, and trips less than 30 miles, which have made suburbia what it is today, would actually cut your fuel bill by a factor of more than ten, even if all the electricity were from zero-carbon sources like wind and nuclear power. The extra cost of the vehicle would be paid for in fuel savings in fewer than five years.

I discuss plug-ins in more detail later (in this chapter).

World's Top Energy Economist Warns Peak Oil Threatens Recovery, Urges Immediate Action: "We Have to Leave Oil before Oil Leaves Us"

August 3, 2009

"Oil prices leapt above $70 a barrel Monday in Asia on investor expectations a recovering global economy will boost crude demand," the AP reports.

You might call those investors speculators—if speculation can be based on marketplace reality. The UK's *Independent* opens its interview with Dr. Fatih Birol, the chief economist at the International Energy Agency (IEA):

> Dr. Birol said that the public and many governments appeared to be oblivious to the fact that the oil on which modern civilisation depends is running out far faster than previously predicted and that global production is likely to peak in about 10 years—at least a decade earlier than most governments had estimated.

The warning is doubly worrisome because until the last year or two, the IEA had been a bastion of relatively staid conservatism and hence useless energy prognostication (like the U.S. Energy Information Administration still is). Now the IEA and Birol have joined the fact-based alarmists, warning in its *World Energy Outlook 2008*, "Without a change in policy, the world is on a path for a rise in global temperature of up to 6°C" and proposing aggressive clean energy solutions.

The IEA's work makes clear that for oil to stay significantly below $200 a barrel (and U.S. gasoline to be significantly below $5 a gallon) by 2020 would take a miracle—or rather six miracles. As the *Independent* reports today:

> But the first detailed assessment of more than 800 oil fields in the world, covering three quarters of global reserves, has found that most of the biggest fields have already peaked and that the rate of decline in oil production is now running at nearly twice the pace as calculated just

> two years ago. On top of this, there is a problem of chronic under-investment by oil-producing countries, a feature that is set to result in an "oil crunch" within the next five years which will jeopardise any hope of a recovery from the present global economic recession, he said . . .
>
> The IEA estimates that the decline in oil production in existing fields is now running at 6.7 percent a year compared to the 3.7 percent decline it had estimated in 2007, which it now acknowledges to be wrong . . .
>
> In its first-ever assessment of the world's major oil fields, the IEA concluded that the global energy system was at a crossroads and that consumption of oil was "patently unsustainable," with expected demand far outstripping supply . . . .
>
> Oil production has already peaked in non-Opec countries and the era of cheap oil has come to an end, it warned.

In short, peak oil is nigh.

As a result of this analysis, Birol has gotten very blunt:

> One day we will run out of oil, it is not today or tomorrow, but one day we will run out of oil and we have to leave oil before oil leaves us, and we have to prepare ourselves for that day . . . The earlier we start, the better, because all of our economic and social system is based on oil, so to change from that will take a lot of time and a lot of money and we should take this issue very seriously . . .
>
> The market power of the very few oil-producing countries, mainly in the Middle East, will increase very quickly. They already have about 40 percent share of the oil market and this will increase much more strongly in the future.

If you thought OPEC and the Persian Gulf producers were powerful before, wait until they control most of the oil market and have more than double their current revenues:

> If we see a tightness of the markets, people in the street will see it in terms of higher prices, much higher than we see now. It will have an impact on the economy, definitely, especially if we see this tightness in the markets in the next few years.

So $4 and $5 gasoline—here we come. As Dr. Birol told *The Independent*:

> It will be especially important because the global economy will still be very fragile, very vulnerable. Many people think there will be a recovery

> in a few years' time but it will be a slow recovery and a fragile recovery and we will have the risk that the recovery will be strangled with higher oil prices.

What needs to be done? The only way to stop oil demand from outstripping the peaking of oil production is massive demand destruction, which is itself possible in only two ways. The first way, pursued by the Bush administration, albeit (mostly) unintentionally, is to destroy the global economy. Let's call that the short-term "non-optimal" approach.

But as IEA has noted, we need to find four to six Saudi Arabias of oil. Merrill Lynch also warned in 2009 that non-OPEC production has likely peaked and that "the cumulative decline of global oil production from today could amount to 30 million barrels per day by 2015," as Arabianbusiness.com reported:

> Steep falls in oil production means the world now needed to replace an amount of oil output equivalent to Saudi Arabia's production every two years, Merrill Lynch said in a research report.

A March 2009 McKinsey report concluded, "the potential looms for liquids demand growth to outpace supply creating a new spike in oil prices as soon as 2010 to 2013, depending on the depth of the economic downturn."

The Obama administration certainly understands that "the equivalent to Saudia Arabia's production every two years" can't be found underground.

It can only be found in our grotesquely inefficient oil consumption. Hence they have advanced the most aggressive increase in fuel economy standards proposed in decades, raising new car fuel efficiency standard to 39 mpg by 2016, which is "the biggest step the U.S. government has ever taken to cut CO_2." And China plans tougher fuel standards than we do.

Hence, the administration has launched a massive push toward plug-in hybrid electric vehicles (PHEVs) and pure EVs in the stimulus, the budget, and the climate bill, which is the core oil and climate solution (see below). As has China, which may offer the world's first mass-market plug-in hybrid for $22,000. No surprise, then, that the major car companies are pursuing PHEVs. Ford expects 10 percent to 25 percent of fleet to be electric by 2020,

Toyota plans up to 30,000 plug-ins in 2012, and GM is introducing the 230 mpg Chevy Volt.

Will all this effort be too little, too late to avoid $200 a barrel oil? I think so. But it may get enough technology into the marketplace by, say, 2015 that when we get really desperate and are ready to embrace a WWII-scale deployment strategy, we will have the commercialized solutions we need. That is the best-case scenario right now.

The Car of the Future Is Here: The Prius? Nope. Hydrogen? Forget about It. Plug-in Hybrids Are the Only Way to Drive

January 22, 2008

When is someone going to offer a practical and affordable family car that runs on something other than oil and that sharply reduces both greenhouse gas emissions and your fuel bill? A few weeks ago, I test-drove this mythical car of the future, a plug-in hybrid electric vehicle whose mass production might be only a few years away.

The Extreme Hybrid from AFS Trinity was rolled out last week at the Detroit auto show. It can run 40 miles on electricity before reverting to running efficiently on gasoline like a normal hybrid, such as the Toyota Prius. Because the majority of people drive less than 40 miles a day, that car can replace most weekly gasoline use, even if it is charged only once a day. The fuel cost per mile, while running on electricity, is under one-third the current cost of gasoline. A full overnight charge might cost a dollar. The car accelerates like a cheetah, though quietly.

Time is running out on developing a truly energy-efficient car. Accelerated burning of fossil fuels is bringing us closer to the tipping point of irreversible climate catastrophe. We are likely to peak soon in the production of conventional oil—so gasoline prices are inevitably headed higher in the coming decades. Meanwhile, the cars we build today stay on the road more than fifteen years, so we have no time to waste.

You can buy a flexible fuel vehicle today that runs on 85 percent ethanol and 15 percent gasoline, and you might even be able to find an E85 station in your city. But corn ethanol is far from a desirable alternative fuel. It doesn't significantly reduce greenhouse gas emissions, or your fuel bill. That would require low-carbon ethanol from biomass such as switchgrass, so-called cellulosic ethanol, but the country does not have a single commercial cellulosic ethanol facility. It will probably be at least fifteen years—and possibly twice that long—before we have large volumes of cellulosic biofuels for sale nationwide at an affordable price.

Hydrogen cars are even farther away from being practical. Carbon-free hydrogen is likely to be more expensive than gasoline for a long time. And the cost of building a carbon-free hydrogen fueling infrastructure is several hundreds of billions, if not more than a trillion, dollars.

Only one zero-carbon alternative fuel is substantially cheaper than gasoline: electricity from renewable sources (or nuclear power). Of course, you'd need a car that runs on electricity, and many people have thought that you would need a technological breakthrough, or at least a major advance in battery technology, to make that practical.

But game-changing breakthroughs in the energy sector are rare indeed. One can wait a lifetime for a major new technology that fundamentally alters the way we use energy. That's why the Extreme Hybrid, whose electric technology is available today, is so exciting.

We saw all-electric cars in the 1990s, but they failed for a variety of reasons, as explained in the movie *Who Killed the Electric Car?* One problem is that giving an electric car a 200-mile range requires a lot of batteries, which adds weight, takes up space, and increases cost. Plus, it takes hours to fill one up, so if you run out of juice, you are stuck, making it impractical as a primary family car. Ultimately, it lacked support by the very car companies, like General Motors, that built it in the first place.

Everything changed with the success of hybrid-electric cars like the Prius, which combine a gasoline engine with a battery and electric motor. These hybrids charge the battery with energy regenerated during braking or from the gas engine. They prove that a car combining gas and electric drives can be practical and affordable and even desirable. Some groups have been retrofitting Priuses to make them plug-ins, providing the best of both worlds—acting as an electric car for local trips, but keeping the gas tank and engine for long trips and quick refueling.

The key obstacle to building a practical plug-in hybrid has been the battery. Not only do you need a lot more batteries for a plug-in than for a simple hybrid, you need batteries with substantially different capability. Gasoline hybrids mostly need batteries that can provide a lot of power when necessary—such as for accelerating onto a highway—as opposed to batteries that can store a lot of energy, which is what is required to go relatively long distances after a single charging. Designing a single battery that can store a lot of energy and handle power surges is no easy task, especially when that battery must be compact, affordable, and safe as it constantly cycles through various uses.

The *New York Times* describes the problem using this unintentionally amusing mixed metaphor:

> In fact, the problem in a hybrid is not only how much energy the batteries hold, a quality called energy density, but how fast they can deliver it, called power density. The difference between energy density and power density is like the difference between a wine jug and a peanut butter jar—the containers may have the same capacity, but the size of their openings differ greatly.

Note to *NYT*: When describing a power battery that can deliver energy in short, quick bursts, "peanut butter" is not the best analogy. A shaken bottle of champagne might be better.

Regular hybrids were made practical by the development of the nickel metal hydride battery, due in large part to a government-funded research consortium. The prototype or demonstration hybrids built to date have tended to use the more expensive, but more powerful and compact, lithium-ion batteries popularized by the electronics industry.

Yet discharging a battery too rapidly, especially the current generation of relatively inexpensive lithium-ion batteries such as are found in laptops and cellphones, can damage it, degrading its lifetime. The question has been: When will we have an affordable, safe, compact, and long-lived lithium-ion battery that can deliver both energy (for range) and power (for acceleration) sufficient for a practical car?

The failure to find such a battery is a main reason Toyota and GM have been slow to commit to offering a plug-in to consumers. At the recent Detroit auto show, Toyota said it would offer a plug-in by 2010, although at first only to governments and corporations, not to consumers. GM, which for much of the past year has been promising to deliver the plug-in Chevy Volt in commercial quantities in 2010, has recently said that 2010 remains only a goal—no promises.

So how has AFS Trinity bypassed the need for a new lithium-ion battery? Instead of waiting for a battery that can deliver both energy and power cheaply, it uses current lithium-ion batteries for energy, and then adds something called an ultracapacitor for rapid discharge during acceleration.

Ultracaps have 10 to 100 times the power density of typical batteries, but only one-tenth the energy density, so this is a marriage made in heaven, or at least Silicon Valley. The ultracap is the electrical equivalent of the shaken champagne bottle—although even that analogy is flawed since ultracaps do not just discharge quickly, they also charge quickly. That's another benefit that ultracaps bring to hybrids.

Regular hybrids get much of their efficiency gains from their ability to capture the energy normally lost during braking and convert it to electricity. Current hybrid batteries take up only about half of this electricity, but fast-acting ultracaps can take up much more.

AFS Trinity is not an auto company. It applied its technology to retrofit a Saturn Vue hybrid crossover vehicle, with the help of a leading auto engineering company, Ricardo. It believes that with mass production, an Extreme Hybrid would cost only $9,000 more than an ordinary hybrid, with a payback of the extra cost in fuel savings in less than four years at current gas prices. But that will require the company to find a major auto-manufacturing partner willing to commercialize the vehicle.

Of course, like many small companies with great technology that I've seen over the years, AFS Trinity may not succeed in achieving mass production or meeting its cost targets. That said, I think the importance of the Extreme Hybrid is that it shows there's more than one possible strategy for making a practical plug-in, significantly increasing the chances that someone will succeed.

Plug-ins are not a global warming solution by themselves. The current electric grid is half coal power, so when plug-ins are running on conventional grid power, they cut net greenhouse gas emissions by perhaps one-third, compared to a regular hybrid running on gasoline. They would, however, cut emissions by well over half compared to a conventional vehicle.

The big greenhouse gas savings would come about as plug-ins enable a major transition toward clean electricity and away from petroleum-based fuel, which is getting dirtier every year, as unconventional oil, such as Canadian tar sands, becomes more popular.

Unlike petroleum, electricity is poised to get greener in the future, especially as we fight climate change. Indeed, once we have a national cap on carbon emissions, plug-ins will drive even faster growth of the diverse and growing numbers of carbon-free electricity sources, which include solar photovoltaics, solar thermal electric, wind, geothermal, nuclear and, potentially, coal with carbon capture and storage. By providing distributed energy storage to the grid, plug-ins could make intermittent renewables like wind power (mostly available at night) more cost-effective—and ultimately assist renewables in becoming the nation's primary source of power.

Also, if in a few years you were buying a plug-in hybrid, which might last until 2030, you can safely bet that gasoline prices then are going to be much higher than today's $3 a gallon. So plug-ins will be the best hedge money can buy against oil shocks. Also, given that most early adopters of plug-ins are likely to be environmentally conscious, I would expect many of them to run their hybrids on 100 percent renewable power, making plug-ins a major carbon reducer from the start.

Preventing catastrophic climate change will require the average U.S. car and SUV to have 80 percent to 90 percent lower carbon dioxide emissions by 2050, compared to current vehicles (the same for trucks, airplanes, and ships). Plug-ins could be an essential enabler of such deep reductions. They can easily be made flexible-fuel vehicles, so if low-carbon cellulosic biofuels prove practical and affordable, they can be the primary liquid fuel for longer trips. Absent high-efficiency vehicles like plug-ins, it is unlikely we will have enough spare arable land and water in 2050 for cellulosic biofuels to provide sufficient fuel to achieve such deep reductions across the entire transportation sector.

Another key point is that most of the growth in car use in the coming decades will come in countries where people don't necessarily drive long distances on a regular basis—and don't require large SUVs. I believe that an affordable and purely electric car with a range of 200 miles, even one with 100 miles, will be a successful primary car for most people in most other countries. Plug-ins can help enable that transition by reducing the cost of batteries and electronics.

Plug-ins may not suffer from a problem that has plagued so many alternative fuel vehicles—high initial cost. As I told AFS Trinity, I wouldn't recommend designing the first plug-ins with a 40-mile all-electric range. For many people, including me, that represents a capability they are unlikely to use most days, meaning we would be paying for a lot of unnecessary batteries and other electronics. You could cut the cost of the first plug-ins by thousands of dollars if the cars just had a 20-mile all-electric range.

I expect many early adopters will be able to charge their cars twice: at home during the night and at work during the day. Companies like Google, which aggressively supports the development of plug-ins, will surely give its workers a large incentive to buy a plug-in, powered by a charging station at its headquarters that draws on renewable energy. Utilities will be eager to set up charging stations at public places like malls and parking garages. Yes, you would probably not be able to charge for a couple of hours during peak demand on the hottest summer days, but that would still leave you plenty of opportunity to charge your car during the day as well as the night. That means a plug-in with a 20-mile all-electric range would still allow many commuters to drive 40 miles or more on electricity, again significantly reducing batteries, electronics, and cost.

Also, a utility or other intermediary might lease a plug-in hybrid—or at least its battery—to a consumer or business willing to leave the vehicle connected when it was not on the road and to permit the utility to control when the vehicle's battery was charged (and possibly when it was discharged). This would provide the utility with a new source of revenue and the consumer with a far less expensive car. A related possibility is being pursued by many companies: Plug-in hybrids could be charged at off-peak times and provide power and voltage to the grid when needed. Vehicle owners may be able to get a rebate or revenue stream from electric utilities for this service.

No country has ever delivered a mass-market alternative fuel vehicle without government mandates. Plug-ins will no doubt need initial help, although they probably require less government intervention than other alternative fuels since they don't require an entirely new fueling infrastructure. To spur their development, Brookings Institution scholar and White House veteran David Sandalow recommends that the federal government buy 30,000 plug-ins at an $8,000 premium. He suggests that the government offer an $8,000 consumer tax credit for purchasers of the first million plug-ins, and a $4,000 rebate for purchasers of the second million.

These steps, coupled with a big push toward low-carbon electricity, would speed the day when we have a practical, clean, plug-in hybrid. And that now seems closer than ever.

CHAPTER 5

The Clean Energy New Deal

A CLEAN ENERGY FUTURE IS INEVITABLE. We have a limited supply of fossil fuels, especially oil. Our burning of fossil fuels is destroying a livable climate.

The two key questions are, first, will we voluntarily give up fossil fuels in the next couple of decades, rather than being forced to do so helter-skelter after it is too late to stop the catastrophe? Second, when we do give them up, will the United States be a global leader in creating jobs and exports in clean energy technologies or will we be importing them from Europe, Japan, and the likely clean energy leader in our absence, China.

For more than a quarter century, conservatives have blocked or scaled back efforts by progressives to spend more on clean energy development and deployment. As a result, while we lead the world in virtually every type of clean energy through the early 1980s, now we are playing catch up across the board, even in technologies that we invented, like the solar cell and the efficient lightbulb.

At last we have a president and Congress who get clean energy and global warming. The 2009 economic stimulus bill alone injected nearly $100 billion in federal cleantech investment, which was matched by another $100 billion in private investment. So much has happened in the last year that—though it is not yet enough to avert catastrophe—we can barely comprehend how dramatic a shift it represents.

The Green FDR: Obama's First 100 Days Make—and May Remake—History

April 26, 2009

The media just keeps missing—or messing up—the story of the century.

Future historians will inevitably judge all twenty-first-century presidents on just two issues: global warming and the clean energy transition. If the world doesn't stop catastrophic climate change then all presidents, indeed, all of us, will be seen as failures and rightfully so.

How else could future generations judge us if the United States and the world stay anywhere near our current emissions path, warm most of the inland United States 9°F or more by century's end, with sea levels 4 to 6 feet higher, rising perhaps a half an inch or more a *year*, with the Southwest from Kansas to California a permanent dust bowl, and much of the ocean a hot, acidic dead zone? These impacts could be irreversible for hundreds of years if we don't reverse emissions soon and sharply. This will require an unbroken—and indeed escalating—response by our political leadership throughout this century. The same is true for the very important, but still secondary, issue of avoiding the worst impacts of peak oil.

In that sense, what team Obama has accomplished in its first 100 days is nothing less than an unprecedented reversal of decades of unsustainable national policy forced down the throat of the American public by conservatives. While I will present a longer list later in this section—and welcome your additions—three game-changing accomplishments stand out:

1) **Green Stimulus**: Progressives, Obama keep promise to jumpstart clean energy, economy—conservatives keep promise to jumpstop the future
2) **Sustainable Budget**: The first sustainable budget in U.S. history.

3) **Regulatory breakthrough**: EPA finds carbon pollution a serious danger to Americans' health and welfare requiring regulation.

Obama has clearly demonstrated he has a serious chance to be the first president since FDR to remake the country through his positive vision. Indeed, if Obama is a two-term president, if he achieves even half of what he has set out to, he will likely be remembered as "the green FDR."

As an interesting side note, President Reagan, who is held in some esteem with historians these days, will almost certainly be relegated to a second-tier, if not third-tier, president by the painful dual realities of global warming and peak oil. After all, it was Ronald Reagan who put conservatives strongly and permanently on the pro-pollution, anti-efficiency, anti-clean-energy side, where they remain today. It is Reagan, more than anyone else, who put the GOP on the self-destructively wrong side of scientific reality (though Newt Gingrich is a close second).

Since the establishment media doesn't get global warming—seeing it mostly through the lens of their standard drama- and personality-driven coverage focused on the ephemeral (whether Obama "blinks" on earmarks, or Newt Gingrich faces off vs. Al Gore)—and since establishment historians almost by definition focus on the past, the overwhelming majority of "first 100 days" articles you will read are irrelevant exercises in navelgazing. I won't even bother linking to or debunking the spate of stories in today's *New York Times* or *Washington Post* Sunday sections.

These myopic stories all befit an industry so shortsighted it couldn't even understand the implications for its own future of the Internet revolution it was reporting on. As but one of many painful examples, here is Joe Klein writing in the normally green-savvy *Time*, "Sizing Up Obama's First 100 Days":

> The fate of Obama's first year in office, if not his Administration, will probably be determined by the way he handles four distinct challenges—two in foreign policy and two domestically . . .
>
> And that's the second domestic challenge: the realization that Congress will not give Obama everything he wants. Aides say the President's moments of frustration almost always have to do with Congress. "We

> know that not every wagon makes it across the frontier," says a top Obama adviser. "But we're not willing to decide yet which wagons are going to make it and which aren't." In fact, that decision seems more and more apparent: Congress is unlikely to pass the linchpin of Obama's alternative-energy initiative—a cap-and-trade program for carbon emissions to combat global warming and tilt the market toward energy independence but that would also raise energy prices in the midst of a recession.
>
> "The wagon that needs to get through is health care," says a second Obama adviser, picking up the metaphor.

Note the utter lack of knowledge or interest in the substance of the global warming problem. Note the backward view of the core issue: Cap-and-trade is not the linchpin of Obama's alternative energy initiative—it is alternative energy that is the linchpin of Obama's effort to avert catastrophic global warming.

Note that Klein, another status-quo establishment journalist like David Broder and Evan Thomas, parrots the standard conservative talking point that Obama wants to "raise energy prices in the midst of a recession," when the cap doesn't even kick in until 2012, and hardly bites for another five years after that. Seriously guys, can you think for yourselves?

Note the selective quoting meant to imply that Obama is ready to throw cap-and-trade overboard to save health care, when anybody who actually listens to *any* of Obama's major speeches would know how nonsensical that view is. For instance, in a big April 14 speech focused on the economy, Obama reaffirmed his commitment to a clean energy economy and strong climate bill: "The only way to truly spark this transformation is through a gradual, market-based cap on carbon pollution"

Obama gets global warming. The media doesn't.

Anyway, let's move from the out-of-touch chattering class to a class in green leadership. How has Obama jumpstarted the one true task of every U.S. president of the twenty-first century—preserving the health and welfare of the next 100 billion people to walk the Earth?

Here is a partial list of what Obama has achieved in his first 100 days, laying the groundwork for him becoming the Green FDR:

1) Obama began the process of blocking the vast majority of new coal plants. The EPA has reversed the Bush EPA's effort to ignore the Supreme Court decision that determined carbon dioxide was a pollutant (and hence that CO_2 emissions from new coal-fired power plants needed regulating) and initiated the process of regulating greenhouse gases for the first time in U.S. history.

2) He began the process of dramatically increasing the efficiency of our vehicles, by ordering EPA to quickly give California and a dozen other states the right to put in place tough emissions requirements for tailpipe emissions of greenhouse gases—and by ordering the Department of Transportation to quickly issue and phase-in tougher fuel economy standards to comply with the 2007 Energy Bill, the first overhaul of the nation's fuel efficiency standards in more than three decades.

3) He appointed a first-rate cabinet and then unleashed them to start inconvenient-truth telling to the public after eight years of administration denial and muzzling of U.S. scientists (Energy Secretary Steven Chu has said "Wake up . . . we're looking at a scenario where there's no more agriculture in California," and "This is a real economic disaster in the making for our children, for your children").

4) In every single major speech, he has focused on the urgent need for the clean energy transition, for a price for carbon (cap-and-trade and "closing the carbon loophole"), and the unsustainability of our current economic system (Obama gets that our economy has been a Ponzi scheme: "The choice we face is not between saving our environment and saving our economy. The choice we face is between prosperity and decline.")

5) He signed into law the tax credits needed to achieve his ambitious goal of 1 million plug-in hybrids by 2015—the key alternative fuel vehicle strategy needed to avert the worst consequences of three decades of successful conservative efforts to stop this country from dealing with the energy/economic security threat of rising dependence on imported oil and the inevitably grim impacts of peak oil. He also enacted into law $2 billion in grants and loans for R&D into advanced vehicle batteries, a tenfold increase over current funding. Plug-ins and electric cars, of course, are a core climate solution, since electric drives are more efficient, easily powered by carbon-free energy and indeed far cheaper to

operate per mile than gasoline, even when running on renewable power. In the longer term, plug-ins and electric cars can also help enable the full renewable revolution.

6) He signed into law a massive investment in mass transit and train travel—and laid out an aggressive vision for a high-speed rail network. The 70 percent boost in funding is a crucial effort needed to prepare this country for a time when air travel simply becomes too expensive for most people (and then a slightly later time when air travel is seen as simply too destructive of a livable climate)—a time not very far away—one that the vast majority of readers of this blog will live to see.

7) He signed into law the tax credits needed to meet his ambitious goal of doubling renewables in his first term.

8) He signed into law the funding needed to jumpstart a twenty-first-century smart grid that is critical to enable the renewable energy, energy efficiency, and plug-in hybrid revolution. He also made what may be his most important appointment, Jon Wellinghoff for Energy Commission Chief, who understands we may not ever need another new coal or nuclear plant and who has already begun jumpstarting the new, green grid.

9) He signed into law the single biggest investment in the deployment of energy-efficient technology in U.S. history, along with strong incentives for state governments to fix their inefficiency-promoting utility regulations.

10) For the first time in three decades, he more than doubled the annual budget for advanced energy efficiency, renewable energy, and low-carbon technology after Reagan slashed federal efficiency and renewables investments 80 percent to 90 percent, which launched decades of vehement ideological opposition to clean tech by even so-called moderate and maverick conservatives.

11) He put forward the first sustainable budget in U.S. history, one that invests in clean energy, included cap-and-trade revenue, and seeks repeal of fossil industry subsidies. Yes, he made a serious tactical mistake by tentatively pursuing the possibility of trying to pass a climate bill through reconciliation, which allowed conservatives to score some meaningless tactical political victories and thereby confuse the media into thinking Obama was himself not serious about this issue. In fact his budget and everything he has done as president shows the reverse is true, that he understands the fate of his presidency and the health and

> well-being of the American public rests on his success in passing serious energy and climate legislation.

Years from now, this may well be remembered as the time that progressives, led by Obama, began the climate-saving transition to a sustainable low-carbon economy built around green jobs.

Of course, it's entirely possible that this history-making first 100 days won't remake history. It's more than possible that we won't stop catastrophic warming. But if we don't stop the hundreds of years of misery, of Hell and high water, that will almost certainly be because the conservative movement threw their entire weight behind humanity's self-destruction (see chapters 7, 8, and 9)—because conservatives in both chambers refuse to conserve anything, including a livable climate, and willingly sacrificed the health and well-being of the next fifty generations of Americans for their ideology.

But even if we fail to stop the catastrophe, there is no escape from Americans, indeed, all humans, ultimately having a low-carbon, low-oil, low-water, low-natural-capital lifestyle. And thus the vast majority of Obama's initiatives will be recognized by future generations and future historians as the point at which the U.S. government embraced the inevitable and started down the sustainable path that presidents either chose to embrace voluntarily in time to avoid the worst impacts or were forced to embrace by the collapse of the global Ponzi scheme.

Obama is the first president in history to articulate both the why and how of the sustainable vision—and to actively, indeed aggressively, pursue its enactment. And that is why he is likely to be remembered as the green FDR.

Have China and the United States Been Holding Talks Aimed at a Climate Deal This Fall?

May 19, 2009

For those of us who believe that maintaining a livable climate pretty much depends on a U.S.-China deal on greenhouse gas emissions, the *Guardian*'s story Monday was a bombshell:

> **China and United States Held Secret Talks on Climate Change Deal**
>
> - Negotiations began in final months of Bush administration
> - Obama could seal the accord on cutting emissions by autumn

But was the story true? Turns out I know one of the key players:

> "My sense is that we are now working towards something in the fall," said Bill Chandler, director of the energy and climate programme at the Carnegie Endowment for International Peace, and the driving force behind the talks. "It will be serious. It will be substantive, and it will happen."

I've known Bill since my DOE days, so I called him to get the scoop. He says the story is mostly true—and thus a true potential breakthrough that may well lead to a major announcement in the fall—but it has inaccuracies, including the nature of the deal being discussed. Let me try to separate fact from hype and examine what China might be willing to commit to (assuming we makes serious commitments, too, á la Waxman-Markey).

Bill explained that the talks were not secret, but were merely off the record. Indeed, Carnegie had written about the talks back in March.

The *Guardian* is correct that "The first communications, in the autumn of 2007, were initiated by the Chinese. Xie Zhenhua, the vice-chairman of the National Development and Reform Commission, the country's central economic planning body." The *Guardian* puffs up the role of the Bushies in this talk, but in fact these talks were designed to get around the Bush administration intransigence on the issue.

Bill told me that Minister Xie "said that the Chinese government took the science of climate change seriously" and wanted help figuring out who they should talk to given the pending U.S. election and the fact that the "Bush administration wasn't doing anything." China wanted to have off-the-record meetings with "potentially influential people" in the U.S. presidential campaigns so that official negotiations could "hit the ground running" once a new administration was in power. China wanted to achieve an understanding with the United States before the big international climate negotiations meeting in Copenhagen this December—hence the desire to start the dialogue before January 20 of this year.

Bill told me, "I personally have the opinion that a deal is in reach." That, of course, begs the question of what the deal is, which I discussed with Bill at length. Here is where the *Guardian* got the story quite wrong:

> Chandler said he and Holdren drew up a three-point memo that envisaged:
>
> - Using existing technologies to produce a 20 percent cut in carbon emissions by 2010.
> - Cooperating on new technology including carbon capture and storage and fuel efficiency for cars.
> - The United States and China signing up to a global climate change deal in Copenhagen.
>
> "We sent it to Xie and he said he agreed," said Chandler.

The reporter, like many people, has confused an absolute reduction in carbon emissions with efforts to reduce energy and carbon intensity (per dollar of GDP). China had previously announced a goal to cut energy intensity (energy per GDP) by 20 percent by 2010, which from my perspective was not much of a target, since it didn't stop an accelerated use of coal even if they had met their annual efficiency targets, which they didn't.

The priority is to get China off its staggeringly unsustainable trajectory of coal consumption and carbon dioxide emissions growth of the first part of this decade. If they don't, they can single-handedly finish off the climate no matter what we and the other rich countries do. Now the good news is that China has an excellent track record on achieving gains in energy efficiency and has begun to ramp up its efficiency efforts and aggressively expand its carbon-free electricity targets (recently committing, for instance, to triple its wind goal to 100,000 MW by 2020).

China is almost certainly not going to agree to a hard cap this year. And it is not news that China has been contemplating a strong carbon *intensity* (CO_2 per GDP) target. But it would be news if, as Bill says, they are willing to publicly agree to aggressive and enforceable energy efficiency and carbon intensity targets, including a 50 percent carbon intensity cut by 2020.

Bill also believes that "a hard cap in emissions is possible" by China for 2025 with a major inflection point around 2020. He points out that Jiang Kejun, director of the Energy Research Institute, one of China's leading policy thinktanks, has been delivering a very strong presentation about how China could quickly move toward a low-carbon economy. In March 2009 Jiang gave a presentation, "Low-Carbon Scenario up to 2050 for China," to the 18th Asia-Pacific seminar on climate change, in which he lays out how China could meet a low-carbon (LC) scenario that flattens out their emissions trajectory by 2020.

But now we are getting way ahead of ourselves. Before the cap, China needs to show itself and the world it can achieve a sharp slowdown in emissions growth.

Thus, first, we need to get a deal with China that includes some serious, verifiable commitments. Needless to say, that presupposes the United States itself can pass a law like the Waxman-Markey climate and clean energy bill that puts us on the path to sharply reduce our emissions.

The U.S. House of Representatives Approves Landmark (Bipartisan!) Climate Bill, 219–212. Waxman-Markey Would Complete America's Transition to a Clean Energy Economy, Which Started with the Stimulus Bill

June 26, 2009

My Salon piece, "One Brief Shining Moment for Clean Energy," is up. We do need to savor moments like these, since, as I note in that article, given modern conservative ideology, which is 100 percent anti-conservation, "the country can only contemplate serious environmental legislation when we have the unique constellation of a Democratic president and [large] Democratic majorities in both houses, an occurrence far rarer than a total eclipse of the sun."

Every journey of a 1,000 miles begins with a single step—including stopping human-caused global warming at "safe levels," as close as possible to 2°C.

This bill would complete America's transition to a clean energy economy, which was begun in the stimulus. In April, the U.S. Energy Information Administration revised its long-term forecast and now projects wind power will be 5 percent of U.S. electricity in 2012 and all renewables will be at 14 percent, thanks to the stimulus. Within four decades, the vast majority of American's carbon dioxide emissions and fossil fuel consumption will be replaced by a suite of new technologies (discussed in chapters 3 and 4).

This bill makes possible an international deal in Copenhagen this December—as well as a bilateral deal with China, hopefully sooner. Had the bill failed, the chance of humanity avoiding catastrophic climate change would be all but eliminated. As Nobel Prize winner Al Gore wrote earlier today, there was no "backup plan" to Waxman-Markey. In this post, I will revise and extend the post I wrote after the bill passed the Energy and Commerce Committee.

Many people have asked me how I can reconcile my climate science realism, which demands far stronger action than the Waxman-Markey bill requires, and my climate politics realism, which has led me to strongly advocate passage of this flawed bill.

The short answer is that Waxman-Markey is the only game in town. If it fails, I see little real chance of stabilizing anywhere near 350 to 450 ppm since serious U.S. action would certainly be off the table for years, the effort to jumpstart the clean energy economy in this country would stall, the international negotiating process would fall apart, and any chance of a deal with China would be dead. Warming of 5°C (9°F) or more by century's end would be all but inevitable, with 850 to 1000+ ppm. If Waxman-Markey becomes law, then I see perhaps a one in three chance of averting catastrophe—not high enough, but not near zero.

For climate-politics realists, the vote today is a staggering achievement. Today was the first time the U.S. House of Representatives has ever voted on climate legislation. This country hasn't enacted a major economy-wide clean air bill since the Clean Air Act amendments of 1990. And that bill had a cap-and-trade system where 97 percent of the permits were given to polluters. And it focused on direct, obvious, short-term health threats to Americans. And that was a long time ago in a galaxy far, far away, when the entire Republican establishment wasn't dead set against any government-led effort to reduce pollution.

Yet Waxman-Markey did get eight Republican votes, which is eight more than the stimulus bill got! This bill needed Republican votes, which will also be true in the Senate. The closeness of the House vote—with 44 Dems voting No—makes clear that the really hard work is yet to come.

And for those who say this doesn't do enough—I agree 100 percent. But then the original Clean Air Act didn't do enough. And the 1987 Montréal protocol would not have stopped concentrations of ozone depleting substances from rising and thus would not have saved the ozone layer. But it began a process and established a framework that, like the CAA, could be strengthened over time as the science warranted. The painful reality of climate change is going to become increasingly obvious in the coming years, and strengthening is inevitable.

In an earlier post, I discussed the myriad forces lined up against serious climate action. I won't repeat that here, but instead want to excerpt something that David Corn wrote for *Mother Jones*, which states the climate-politics realist position very well—a position you might not associate with Corn and *MJ*:

> So should progressives back this not-a-full-loaf bill? Über-blogger Matt Yglesias offers this hard-headed guidance: follow the Waxman. Citing a recent Charles Homans profile of Waxman, he writes:
>
> There's simply nobody else in Congress whose record of progressive legislative accomplishments can hold a candle to Waxman's. When you draw intersecting curves of "what needs to be done" and "what can realistically be done," Waxman has time and again put himself at the intersection, and I think it involves a fair amount of hubris to think that you know better than him what the best feasible legislative outcome is.

I would add Representative Ed Markey to this equation. For decades, Markey has been a passionate champion of environmental and clean energy causes. A few months ago, he complained to me about Washington's inability to address the threat of climate change. Like Waxman, he gives a damn about this and truly wants to pass the toughest bill possible.

Enviros can decide for themselves how much compromise to accept. Ultimately, our political system may not at this time—even with President Barack Obama at the helm—be able to handle the full truth about climate change and act accordingly. But it's hard to second-guess Markey and Waxman. If they are cutting deals, they are doing what they reluctantly need to do, not what they want.

It will be a staggering achievement if, in six to nine months, an energy and climate bill that looks something like Waxman-Markey is signed into law by President Obama.

From the perspective of political realism, though, it will be a great challenge just to stop this bill from being weakened as it winds itself through the House and especially the Senate. Indeed, it should be strengthened. That is the hard task ahead.

From the perspective of climate science realists, the bill has two flaws, one of which is very serious. And I don't mean the allocations for big polluters. I know many of my readers disagree, but I just don't think that the allocation undermines the goals of the bill at all, and in fact are a perfectly reasonable way of satisfying political needs while preventing windfalls for polluters and preserving prices. Harvard economist and cap-and-trade expert Robert Stavins has written that "The appropriate characterization of the Waxman-Markey allocation is that more than 80 percent of the value of allowances go to consumers and public purposes, and less than 20 percent to private industry."

The first flaw is the 2 billion offsets that polluters can potentially use instead of their own emissions reductions. I have previously explained why I am far less worried about domestic offsets. In a regulated market with a cap, many of the domestic offsets will represent real reductions of US greenhouse gas emissions, and the total supply of cheap domestic offsets will be limited. I have also explained why I do not believe the international offsets threaten the overall integrity of the bill. The key point is that last year, the entire international offsets market utilized by the Europeans was 82 million tons with an average price of $25/ton (and about half of those tons were questionable, low-cost tons from reducing HFC refrigerants from China that won't be available by 2012). If the United States comes into the international offsets market even in a modest way, the price will certainly be higher than that, especially if we work to improve offset quality as the bill demands. Still, I'd love the Senate to improve the bill by sunsetting the offsets, phasing them out over time.

The bottom line is that the vast amounts of moderate-cost near-term domestic emissions reductions strategies—energy efficiency, conservation, replacing coal power with natural gas-fired power, wind power, biomass cofiring, recycled energy and so on—will be available at a net cost of $20/ton or less (in quantity) in 2020, which is considerably lower than the likely cost of international offsets.

And that brings us directly to the second and far graver flaw—the 2020 target is too weak. Given the lost eight years of the Bush administration, it was inevitable that a bill that doesn't even impose a cap until 2012 could not have the same 2020 target (compared to 1990 levels) that the Europeans are considering (a 20 percent to 30 percent reduction).

That means we're going to build too much polluting crap in the next decade. That means we'll have to go back and unbuild it at some point. More expensive, sure, than doing it right the first time, but no more difficult than deploying the dozen or so accelerated stabilization wedges globally in three to four decades needed to beat 450 ppm.

For me, a two-term President Obama (together with the next three Congresses) cannot solve the global warming problem, but can create the conditions that allow the next couple of presidents to do what is needed. Or he can be thwarted, making it all but impossible for future presidents.

The only hope for stabilizing at 350 to 450 ppm is a WWII-scale and WWII-style effort as I have said many times. And that implies a level of desperation we don't have now, a desperation that will come as the planet heats up, yet more ice melts faster than anyone expected, and yet more droughts, heat waves, and superstorms are inflicted on us. When we have that desperation, probably in the 2020s, we'll want to already have:

- substantially dropped below the business-as-usual emissions path
- started every major business planning for much deeper reductions
- goosed the cleantech venture and financing community
- put in place the entire framework for U.S. climate regulations
- accelerated many tens of gigawatts of different types of low-carbon energy into the marketplace
- put billions into developing advanced low-carbon technology
- started building out the smart, green grid of the twenty-first century
- trained and created millions of clean energy jobs
- negotiated a working international climate regime
- brought China into the process

This bill is crucial to achieving all of those vital goals.

Kudos to Nancy Pelosi and Henry Waxman and Ed Markey and President Obama—and a great many other progressive politicians and advocates—for making this historic moment happen.

Enjoy the weekend. The really hard work—Senate passage—is next (see afterword).

CHAPTER 6

The Bush-Cheney Reign of Error

NO TWO PEOPLE IN THE HISTORY OF this country or indeed the world have done more to ensure that humanity will not act in time to stop catastrophic global warming than President George W. Bush and Vice President Dick Cheney.

If we don't preserve a livable climate, future generations will judge them far more harshly than we do today. As a blogger, it was sometimes hard to know whether to cry or laugh at their absurdities. So I did both.

Climate Progress Person of the Year for 2007

December 20, 2007

Until last week, this long-beloved annual tradition of Climate Progress seemed to be a lock for one person—Nobel laureate, itinerant educator, and media superstar Al Gore. Sadly, he only makes first runner up this year. Similar to *Time* magazine, our Person of the Year is awarded to the person or group who "for better or for worse . . . has done the *most* to influence the events of the year" in the climate arena.

By singlehandedly stopping any international action on climate at Bali, by stopping California from regulating tailpipe greenhouse gas emissions, by forcing Congress to drop almost all non-oil-related provisions to cut GHGs from the energy bill—all in one week!—one man proved his unchallenged high-impact misleadership on the issue of the century: George Bush.*

In a related story, the FHS (Future Historians Society), having previously named Bush the Worst President in American History, awarded him one of their rare Worst Leaders of All Time Awards, alongside such notables as Neville Chamberlain and Nero, for his tireless efforts to destroy the health and well-being of the next fifty generations.

Bush spokesperson, Dana Perino, said the President always believed he deserved as much recognition for his global warming efforts as Al Gore.

---

* Note to future historians: Curiously, Gore seemed to have gotten more actual votes than Bush did for the honor, but the judges awarded it to Bush anyway.

History Won't Warm to "W"

January 16, 2007

President George W. Bush believes history will end up judging him favorably. He compares himself to Harry Truman who left office unpopular in large part because of a difficult war on the Korean peninsula but who is now admired by historians. President Bush suffers from an unpopular war, too, yet absent a dramatic reversal in President Bush's climate policies—never mind Iraq—it's a good bet that neither historians nor future generations of Americans will ever warm to President Bush.

Predicting the long-term consequences of the president's misguided and mismanaged invasion of Iraq is impossible. But it is not at all difficult to detail the suffering that humanity faces because of global warming. If the United States "stays the course" with President Bush's non-interventionist climate policies over the next decade, then by the third decade of this century all of American life—politics, international relations, our homes, our jobs, our industries, the kind of cars we drive—will be forever transformed.

"There can no longer be genuine doubt that human-made gases are the dominant cause of observed warming," explained James Hansen, director of NASA's Goddard Institute for Space Studies, in April 2005. Worse, the great ice sheets atop and anchoring our planet are melting much faster than most scientists thought even a few years ago. John Holdren, president of the American Association for the Advancement of Science, warns that sea levels could rise 7 feet or more by the end of this century.

Our nation, along with the rest of the world, will face suffering that dwarfs the aftermath of Hurricane Katrina. If Bush's policies continue for another decade, then government intervention in our lives on a scale not seen since World War II is probably unavoidable. Historians living through such times will render a harsh judgment on a leader of the wealthiest nation on the planet who not only ignored the warnings about global warming but actively worked to block every effort to avoid or minimize the risk too.

Bush has ignored the wealth of scientific information that makes clear human-induced global warming will not be on the mild side, and may well be catastrophic. Unfazed by the evidence, the president refused early in his presidency to join all other industrialized nations save Australia in action under the Kyoto Accord to reduce emissions of heat-trapping greenhouse gases. Since then, senior administration officials have slammed the door on any serious action in Bush's second term and opposed actions by individual states. Along the way, the Bush White House has stymied international efforts that could have actually led to global reductions in time to avoid the worst.

The president's track record is truly awful. The Kyoto Accord required only a 7 percent cut in carbon dioxide emissions below 1990 levels by 2010, yet he rejected Kyoto, saying it would do irreparable harm to the economy. Then, after intense lobbying by the White House, Congress rejected a still weaker cap that did not actually require an absolute emissions reduction.

Instead of dealing with the threat posed by climate change, the administration has systematically worked to suppress the truth about it. The Government Accountability Office concluded in April 2005 that the Bush climate program "lacks a major component required by law: periodic assessments of how rising temperatures may affect people and the environment." The White House even hired a former lobbyist for the American Petroleum Institute to censor scientific conclusions in government climate reports.

Fortunately, global warming is not inevitable. The president could burnish his historical legacy by embracing serious climate change policies in his 2007 State of the Union speech later this month. If he doesn't, his successors in the Oval Office must slash greenhouse gas emissions such as carbon dioxide by 2050 to 60 percent below 1990 levels. British Prime Minister Tony Blair has vowed the UK will do just that. And Governor Arnold Schwarzenegger of California has committed his state to even deeper reductions.

An unprecedented national and international effort is required to cut greenhouse gas emissions as sharply as possible without crushing economic growth. In addition, unprecedented efforts must be taken to plan for the worst effects of global warming—sea level rises, massive storm surges, widespread heatwaves and drought, and the spread of tropical diseases.

Consider the planning needed just to relocate the hundreds of millions of people who live near the world's shorelines. Imagine the engineering that will be required to build massive dikes around Manhattan, Miami, Houston, San Francisco, and dozens of the world's greatest cities.

The next president of the United States, together with Congress, must quickly repudiate the Bush administration's climate policies across-the-board and embrace an aggressive strategy to bring the myriad tools of government together to address the problem.

These policies include a cap on carbon emissions, a serious increase in fuel economy standards for cars and light trucks, and at least a doubling of federal R&D for clean energy. If Bush were to reverse course and adopt it, such a policy reversal wouldn't spare him the harsh judgment of history, but it would spare the country the worst consequences of his mistakes. Few predictions of the future can be made with such high certainty.

Has Anyone in U.S. History Made More Americans Less Safe than Dick Cheney?

May 11, 2009

People Who Live in Greenhouses Shouldn't Throw Stones

Back in March, former vice president Dick Cheney said Americans are "less safe" now thanks to President Barack Obama and his policies. He repeated and expanded on the charge yesterday on *Face the Nation.*

Let's set aside the fact that if a president's actions and policies in his first 100 days make him 100 percent responsible for any attack on this nation, then Bush and Cheney are 100 percent responsible for 9/11.

Terrorism is a real threat to Americans. But it pales in comparison to the scale and scope of the threat posed by unrestricted emissions of greenhouse gases. In the words of IPCC head R. K. Pachauri—who was essentially handpicked by Cheney himself to replace the "alarmist" Bob Watson:

> The cities, power plants and factories we build in the next seven years will shape our climate in mid-century. We have to act now to price carbon and create incentives to change the way we use energy and spread technology—and thereby avert nothing less than an existential threat to civilization.

As the über-centrist (and formerly left-leaning) Brookings Institution put it in a pre-election op-ed:

> Today's adults, even if they will not be around at mid-century, must think about the fate of their children and grandchildren [who] are among nearly 75 million Americans—and 2.2 billion people worldwide—younger than 18. That generation will be in its 40s or 50s when one of two things happens: Either the temperature of the planet warms

> more than 4.5 degrees and vast regions slide toward being uninhabitable, or the wisdom of the next president and his fellow leaders around the world pays off in the ultimate reward—survival.

Global warming is the only true preventable existential threat to the health and well-being—the security and safety—of Americans.

So I repeat the headline question—has anyone in U.S. history made more Americans less safe than Dick Cheney?

Remember, President Bush campaigned on regulating carbon dioxide emissions from electric utilities. Dick Cheney is the person who killed that. Indeed, it is doubtful that Bush had particularly strong opinions on any major energy or environmental issue. Cheney, after all, is the one who put together Bush's entire energy plan.

Cheney led the effort to block all EPA action on climate and censor U.S. scientists from even telling the American public about the dangers posed by global warming as the Center for American Progress Action Fund detailed in a report:

> Last October, Dr. Julie Gerberding, director of the Centers for Disease Control and Prevention (CDC), testified before the Senate Environment and Public Works Committee about the "Human Impacts of Global Warming." Gerberding told the committee that global warming "is anticipated to have a broad range of impacts on the health of Americans," but she gave few specifics, instead focusing on the CDC's current preparation plans. Soon after Gerberding delivered her testimony, CDC officials revealed that the White House had "eviscerated" her testimony by editing it down from 14 pages to four. . . . In a letter responding to questions by Sen. Barbara Boxer (D-CA) yesterday, former EPA official Jason Burnett revealed that Vice President Dick Cheney's office and the Council on Environmental Quality pushed to "remove from the testimony any discussion of the human health consequences of climate change."
>
> **CHENEY'S MALIGN ENVIRONMENTAL INFLUENCE:** In his letter to Boxer, Burnett revealed that Cheney's office had also objected in January to congressional testimony by EPA administrator Stephen Johnson that "greenhouse gas emissions harm the environment." According to Burnett, an official in Cheney's office "called to tell me that his office wanted the language changed." Such actions are not unusual for Cheney.

> Since taking office, he has taken "a decisive role to undercut long-standing environmental regulations for the benefit of business" while undermining any real action to combat climate change. In December, after Johnson "answered the pleas of industry executives" by announcing his decision to deny California the right to regulate greenhouse gases from vehicles, it was revealed that executives from the auto industry had appealed directly to Cheney. EPA staffers told the *Los Angeles Times* that Johnson "made his decision" only after Cheney met with the executives. Since February 2007, Cheney has quietly maneuvered to exert increased control over environmental policy by federal agencies—particularly the regulations on greenhouse gas emissions.

One could write an entire book on Cheney's singlehanded efforts to destroy a livable climate for your children, grandchildren, and the next fifty generations of Americans. I suspect someone will. The *Washington Post* has already written a long story about his role promoting pollution: "The vice president has intervened in many cases to undercut long-standing environmental rules for the benefit of business."

Cheney is too old to see the worst of what his policies will do. And CBS's Bob Schieffer, host of *Face the Nation*, never seriously challenged any of his assertions.

But history will not be kind to Cheney and Bush. Assuming that we don't avert catastrophic global warming, they will be seen as two of the worst leaders in U.S. history—a judgment some are already issuing. Saleem Huq, a lead author of the Intergovernmental Panel on Climate Change's 2007 report on adaptation, said in December 2008, "Bush will go down in history as possibly a person who has doomed the planet."

Make that, "Bush and Cheney."

Bush Launches Unendangered Species List

April 1, 2009

In a surprise move, the Bush Administration today replaced the Endangered Species List with the Unendangered Species List.

In a news conference, the President said, "I have been advisored that my climate, air, and water policies now threaten most living things. Thorny, the Interior guy—I like to call him the Exterior guy, heh, heh, heh—anyway, Thorny says that compiling the old list would have added $1 billion to the deficit over the next ten years."

Secretary Kempthorne introduced the new list saying, "We just felt it would be a lot easier to identify the species not endangered by our policies."

The Secretary's staff handed out a single page containing the entire list, which includes *Rattus rattus*, kudzu and *Toxicodendron radicans*, *Blattella germanica*, *Plasmodium falciparum* and *Plasmodium vivax*, *endroctonus ponderosae*, African trypanosomiasis and tse-tse fly, *tubercle bacillus*, the order *Scorpiones*, the algae supergroups, and the entire Cactaceae and mosquito families.

Bush then added, "That Thorny—he's like some sort of Noah guy."

Introducing the Bipolar Bear

In a related story, the president called the winner of the "Rename the Polar Bear" contest, Dr. Sara Bellum of Nome, Alaska. As reported earlier, rather than attempting to protect the polar bears' Arctic habitat, which is expected to be ice-free by 2020, the Interior Department held a contest to simply give *Ursus maritimus* a new name.

Dr. Bellum, a taxidermist and practicing psychiatrist, explained her winning entry, "I noticed the bears were getting very sad and tired in the summer when the sea ice melted and they had to spend more and more effort catching fewer and fewer seals. But then come the fall they began moving inland, frantically eating everything in sight, rummaging through garbage and attacking people, which, perversely, seemed to make them quite happy, at least for a while. So that's where I came up with the name. Bipolar bear. *Ursus manic-depressus.* I never expected a call from the president. He asked if he could call me 'brainy.' Like I haven't heard that one before."

The first runner up was Danish mathematician, Bjørn Lomborg, with his entry "Ralop Bear." Lomborg, a well known global warming delayer and practicing taxidermist, has long argued that polar bears would rapidly evolve backwards toward their brown bear ancestors from whom they diverged tens of thousands of years ago. Bjørn, whose name coincidentally means bear, could not be reached for comment, but a post on his blog explained, "Ralop—it's 'Polar' backwards. Get it?"

According to Interior staffers, *Ursus manic-depressus* is the only large mammal on the Unendangered Species List.

Q: Why Does Sarah Palin Wear a Polar Bear Pin?

October 13, 2008

A1: She wants to help people remember what they looked like before her policies render them extinct.
A2: She likes sticking it to the bears.
A3. She couldn't find a wolf-cub pin.

Alaska Governor Sarah Palin is famously fighting the Bush administration liberals who designated the polar bears a threatened species, and her global warming denial, if enshrined into law, would finish off the bear's habitat (see blog post "McCain VP Palin Is a Global-Warming-Denying, Polar-Bear-Dissing, Pat Buchanan Acolyte").

Yet she wears a polar bear pin off her right shoulder.

Notwithstanding Palin's faux pas of wearing white (with a white pin!) after Labor Day, her pin is clearly missing either a bull's-eye or a red circle with a slash through it. I wonder what "Polar Bears against Palin" think of this.

Given that eBay founder Meg Whitman is a McCain supporter, I suppose it is no surprise that you can buy the same pin Palin wears on eBay:

> Replica of Sarah Palin's polar bear lapel pin which she wore during several interviews. This pin is handfinished, made of a ceramic material, and has a very durable finish with push pin back. This pin will show your support for Sarah Palin in a classy way. Dimensions—1" × 2"; Weight—less than 1 oz.

Only $14.99 plus $3.50 shipping. As my daddy used to say, it'd be a steal at half that price.

CHAPTER 7

The Right-Wing Disinformation Machine

If the science of the last few years and the painfully obvious reality of a changing climate haven't persuaded the conservative movement of the dire nature of human-caused global warming, I can't imagine what chain of catastrophes would. We've already had record-breaking droughts, heat waves, wildfires, deluges, super storms, and flooding at home and abroad—just as climate science predicted. And we've had far more loss of ice from Greenland, Antarctica, and the Arctic Sea than anyone expected.

Conservative think tanks, conservative media, conservative pundits, and conservative politicians are the driving force behind the climate disinformation campaign. But while they can stop the country from taking the necessary action to avert catastrophe, they can't actually stop the climate from changing. And so, ultimately, if they destroy the climate, they may destroy themselves.

I can't see how the conservative movement as it now exists could possibly survive having been responsible for ushering in decades if not centuries of untold misery and the intrusive government that such hardship and scarcity will entail. If you don't like limited government, you're really going to hate what the government will have to do to adapt to 9°F or more warming, rapid sea level rise, and dust-bowlification, all lasting for decades if not centuries.

Ultimately, the weird thing about most congressional conservatives is that they don't actually seem to care about conserving anything, not energy, not water, not arable land, not even a livable climate. They want radical change—pushing policies that will remake this planet in unimaginably horrific ways.

Hill Conservatives Reject All Three Climate Strategies and Embrace Rush Limbaugh—What Does That Radicalism Mean for Obama, Progressives, and Humanity?

March 4, 2009

Although surprisingly little-remarked on, the big story of the year so far—at least from the perspective of the fate of America and all humankind—is the hardening of the conservative movement against every possible strategy for dealing with global warming.

If you want to tackle global warming, if you want to avert the unimaginable misery of 5.5° to 7°C warming and 850 to 1000 ppm for the next 100 billion people who walk the planet this millennium, you have only three strategies:

1) Put a serious price on carbon.
2) Spend a gazillion dollars on clean technology development and deployment.
3) Mandate the use of efficient, cleaner technology.*

Now even "moderate" conservatives like John McCain and Judd Gregg have always opposed even the mildest of mandates—requirements that utilities get a fraction of their power from renewable energy. Mandates for renewables and more fuel-efficient cars, of course, can't do much more than stem the rise of emissions, so they are just a piece of the puzzle.

As for the serious carbon price, if it wasn't obvious from last year's Senate debate that congressional conservatives intend to demagogue to death any price for carbon (from a cap or a tax)—it is certainly clear now that they believe it is not only not a losing issue, but a big winning issue. Listen

* And yes, for 450 ppm or lower, you need all three.

to one of the conservative movement's reemerging leaders, Newt "Republicans in Congress turn their lonely eyes to" Gingrich who laid out the demagogic strategy at the Conservative Political Action Conference (CPAC) in DC on Friday:

> Former House Speaker Newt Gingrich charged Friday that the cap-and-trade proposal contained in President Barack Obama's budget amounts to a $640 billion "energy tax" over eight years that will break the new president's pledge not to raise taxes on the middle class.
>
> "Let me get this straight," said Gingrich. "We're not going to raise tax on anybody making under $250,000 a year unless you use electricity. And we are not going to raise taxes on anybody under $250,000 a year unless you buy gasoline. And we are not going to raise taxes on anybody who makes under $250,000 a year unless you buy heating oil. And we're not going to raise taxes on anybody who earns less $250,000 a year unless you use natural gas."
>
> "And I try to think to myself," he added, "even in the left wing of the Democratic Party, where there are some people who are fairly unusual, how many of them don't use heating oil, natural gas, gasoline, or electricity?"

Now what you have to remember about Newt is that while he is a hardcore conservative, some progressives actually quite recently thought he was reasonable on climate issues. There was even an ad of Gingrich and Pelosi sponsored by Al Gore's (!) Alliance for Climate Protection.

Ah, the good old days of faux bipartisanship. I remember them as if they were almost . . . last year.

If you've listened to the news over the last few days, then you've heard this theme pressed by conservatives. I saw a clip of Treasury Secretary Tim Geithner savaged yesterday by conservatives on the House Ways and Means Committee over this. Indeed, the demagoguery started before the hearing:

> "This massive hidden energy tax is going to work its way through every aspect of American life," said Rep. Dave Camp of Michigan, the top Republican on the Ways and Means Committee. "How we light our homes, heat our homes, and pay for the gas in our cars, in every phase of our daily lives, we will be paying higher costs."

So please tell me how many House conservatives are going to vote for a cap. Why should they try to preserve a livable climate for future generations of Americans when they see destroying the climate as the road to their salvation? Congressman James Sensenbrenner (R-WI) said in December 2008 that "Democrats will lose their congressional majorities in 2010 if they try to pass a significant global warming bill along the lines of what President-elect Barack Obama promised during the campaign."

As for the Senate, even holding McCain and the Maine delegation for a serious bill will be a great achievement—but McCain has already said that three senators doesn't count as bipartisanship.

Finally, we have the strategy of spending "real money" on climate—and I'm not talking that $15 billion a year on cleantech that Obama has pledged. No, for those supposedly on the side of saving the climate who aren't big on carbon prices or mandates, who think that it's a much easier political lift to just spend money, Obama's plan is, pardon the expression, Bush league. Advocates of this Apollo program approach want to "get Obama to double, triple, or even quadruple his commitment to the strategic public investments necessary to spark a clean energy economy."†

Yes, I think I've heard the voices of a few conservatives in recent days say that the problem with Obama's plan is that he is spending far, far too little on public investments in clean energy. But then I woke up.

Whatever we do on climate for foreseeable future, it ain't gonna be very bipartisan.

Let me end by reposting what I wrote about in January 2009, what I called "the #2 story of 2008":

Conservatives go all-in on climate denial and delay. While the grim implications of the science and observational data discussed above have become painfully obvious to everyone else, conservatives simply refuse to accept reality. For instance, even though a very warm 2008 makes this the hottest decade in recorded history by far—and even though 2008 was about 0.1°C warmer than the decade of the 1990s as a whole (even with a La Niña–

---

† Pause for laughter, or tears.

fueled cool winter) for some deniers, "2008 was the year man-made global warming was disproved." Seriously.

The entire conservative movement—including pundits, think tanks, and politicians—now appears willing to stake the future of humanity on their willful ignorance.

- Virtually every conservative in the Senate voted against the Boxer-Lieberman-Warner climate bill even though that bill was inadequate to stopping catastrophic warming.
- Grover Norquist asserts that calls to take global warming more seriously will be "cheerfully ignored."
- Sixty-four percent of GOP voters say global warming denier Palin is their top choice for 2012.
- "Several prominent party officials said they believe the GOP's message is fundamentally sound when it comes to energy policy, pointing to that issue as one of the few political bright spots in recent years."
- The Heritage Foundation even opposes energy efficiency.
- The American Enterprise Institute is still denying climate science and trying to delay climate action.
- The Cato Institute believes adapting to catastrophic climate change is cheaper than preventing it from happening in the first place.
- Columns by Charles Krauthammer and George Will and John Tierney have become science-free zones that demand more climate research while inveighing against all serious climate action and against all non-nuclear clean tech.

That's why the professional global warming deniers are winning, especially with GOP voters or rather *only with GOP voters* (see chapter 8).

If the Obama climate dream team is going to lead the nation and the world into a World War II–scale effort to save humanity from self-destruction, they will be waging a difficult two-front war—against the ever-accelerating reality of climate change itself and against the immovable unreality of "anti-science conservatives."

The Real Reason Conservatives Don't Believe in Climate Science

June 1, 2008

Washington Post columnist Charles Krauthammer has written a remarkable, anti-science screed, "Carbon Chastity: The First Commandment of the Church of the Environment." I ended my first post on it asking, "Why does he break faith with so many conservatives and worship at the altar of evolution science, but stick with them on climate denial?" My book, *Hell and High Water*, discusses this general question at length, and offers the answer:

> The answer is that ideology trumps rationality. Most conservatives cannot abide the solution to global warming—strong government regulations and a government-led effort to accelerate clean energy technologies into the market. According to the late Jude Wanniski, Elizabeth Kolbert's *New Yorker* articles [on global warming], did nothing more "than write a long editorial on behalf of government intervention to stamp out carbon dioxide." His villain is not global warming, but is the threat to Americans from government itself.

George Will's review of Michael Crichton's *State of Fear* says: "Crichton's subject is today's fear that global warming will cause catastrophic climate change, a belief now so conventional that it seems to require no supporting data. . . . Various factions have interests—monetary, political, even emotional—in cultivating fears. The fears invariably seem to require more government subservience to environmentalists and more government supervision of our lives."

As the *New York Times*'s Andy Revkin explained about the recent anti-science conference in New York, "The one thing all the attendees seem to share is a deep dislike for mandatory restrictions on greenhouse gases."

What unites these people is their desire to delay or stop action to cut GHGs, not any one particular view on the climate.

It is nearly impossible to win an argument with a conservative or libertarian who hates government-led action. Yes, you can try to point out all the great things the government has done (the Internet, anyone?) and try to point out that they invariably support government-led action for military security, and, of course, government subsidies and regulations to promote energy security, at least as it applies to oil industry and nuclear energy pork.

I have a different argument—if you hate government intrusion into people's lives, you'd better stop catastrophic global warming, because nothing drives a country more toward activist government than scarcity and depravation. Interestingly, Krauthammer understands this point abstractly, but since he has no understanding of climate science, indeed he has no interest in learning about the subject at all, he gets the argument exactly backwards.

If you read Krauthammer's whole climate article, he tries to focus the discussion not on science (which he clearly doesn't understand) but on environmentalism (which he thinks he does). This is a very common denier debating tactic, since deniers are in fact usually debating environmentalists, not scientists, because most scientists don't like to engage in the public arena. He writes:

> Yet on the basis of this speculation, environmental activists, attended by compliant scientists and opportunistic politicians, are advocating radical economic and social regulation. "The largest threat to freedom, democracy, the market economy and prosperity," warns Czech President Vaclav Klaus, "is no longer socialism. It is, instead, the ambitious, arrogant, unscrupulous ideology of environmentalism."

Do you know any serious scientists? "Compliant" is the last word one would ever use to describe them. Indeed, the best way to get famous in science is to be a skeptic, to disprove a widely held belief.

This paragraph restates the heart of why conservatives hate climate science. It requires action by government, which, for conservatives, is the same as socialism (again, except when it comes to government action on behalf of the nuclear and fossil fuel industries, which is good ol' capitalism). Krauthammer continues:

> Environmentalists are Gaia's priests, instructing us in her proper service and casting out those who refuse to genuflect. . . . And having proclaimed the ultimate commandment—carbon chastity—they are preparing the supporting canonical legislation that will tell you how much you can travel, what kind of light you will read by, and at what temperature you may set your bedroom thermostat. . . .
>
> There's no greater social power than the power to ration. And, other than rationing food, there is no greater instrument of social control than rationing energy, the currency of just about everything one does and uses in an advanced society.

Here is where the conservatives have it backward. The solution to global warming—the strategy needed to avoid 450 ppm—does not require rationing food or energy. It primarily requires a government-led strategy to aggressively deploy clean energy technologies, as we've seen. That strategy preserves the energy abundance that has made modern civilization possible.

But if we hold off today on government action that focuses for several decades on preventing catastrophe, we will almost guarantee the need for extreme and intrusive government action in the post-2030 era, perhaps lasting centuries. Only Big Government—which conservatives say they don't want—can relocate millions of citizens, build massive levees, ration crucial resources like water and arable land, mandate harsh and rapid reductions in certain kinds of energy—all of which will be inevitable if we don't act now.

Ironically, Krauthammer is afraid of climate strategies that are "economically ruinous and socially destructive," and says the greatest form of rationing is food rationing. Well, if we follow the talk-much do-little climate strategy of conservatives, then we are all but certain to end up at 1000 ppm (10°F warming) by century's end, and that would be economically ruinous and socially destructive. And long before then, with peak oil prices that we haven't prepared for, hundreds of millions more people to feed and increasing desertification, drought, and loss of inland glaciers, we *will* be rationing food. *And* water.

The scarcity and deprivation of 1000 ppm could last for hundreds of years. Conservatives can't stop 1000 ppm by their anti-science, anti-government rhetoric. But they can prevent progressives and moderates from stopping 1000 ppm by blocking aggressive climate legislation. How ironic—and tragic—it would be if conservatives' short-term quest to avoid a bigger government led to a permanently huge government.

The Washington Post, Abandoning Any Journalistic Standards, Lets George Will Publish for a Third Time Global Warming Lies Debunked on Its Own Pages

April 2, 2009

Fool me once, shame on you, fool me twice, shame on me, fool me three times, shame on the media.

In a move that calls into question the journalistic integrity of the entire *Washington Post* editorial staff—especially editorial page editor, Fred Hiatt, who should be replaced—the newspaper has published a third disinformation-pushing op-ed by George Will, "Climate Change's Dim Bulbs."

The distortions and disinformation in Will's earlier two pieces have been widely criticized and debunked (see my blog post, "In a Blunder Reminiscent of Janet Cooke Scandal, the *Washington Post* Lets George Will Reassert All His Climate Falsehoods Plus Some New Ones").

Indeed, the *Post* published a devastating critique by the World Meteorological Organization on the "misinterpretation of the data and of scientific knowledge" in using WMO data from one year to try to "invalidate the reality of global warming and its effects"—along with an op-ed making the same exact point (see my post, "*Washington Post* Publishes Two Strong Debunkings of George Will's Double Dose of Disinformation").

Why on Earth would the *Washington Post* publish a long letter by the WMO Secretary General last Saturday, explaining how his organizations' work was misused by Will, and then let Will publish this sentence today?

> Reducing carbon emissions supposedly will reverse warming, which is allegedly occurring even though, according to statistics published by the World Meteorological Organization, there has not been a warmer year on record than 1998.

Does the *Washington Post* editorial staff care that Will is playing them for fools? Does the *Post* have any idea whatsoever how amateurish this makes them look, like some high school newspaper?

Let me reprint the bulk of the letter the *Post* published on Saturday:

> Data collected over the past 150 years by the 188 members of the World Meteorological Organization (WMO) through observing networks of tens of thousands of stations on land, at sea, in the air, and from constellations of weather and climate satellites lead to an unequivocal conclusion: The observed increase in global surface temperatures is a manifestation of global warming. Warming has accelerated particularly in the past 20 years.
>
> It is a misinterpretation of the data and of scientific knowledge to point to one year as the warmest on record—as was done in a recent *Post* column [by Will]—and then to extrapolate that cooler subsequent years invalidate the reality of global warming and its effects.
>
> The difference between climate variability and climate change is critical, not just for scientists or those engaging in policy debates about warming. Just as one cold snap does not change the global warming trend, one heat wave does not reinforce it. Since the beginning of the twentieth century, the global average surface temperature has risen 1.33 degrees Fahrenheit.
>
> Evidence of global warming has been documented in widespread decreases in snow cover, sea ice, and glaciers. The 11 warmest years on record occurred in the past 13 years.
>
> While variations occur throughout the temperature record, shorter-term variations do not contradict the overwhelming long-term increase in global surface temperatures since 1850, when reliable meteorological recordkeeping began. Year to year, we may observe in some parts of the world colder or warmer episodes than in other parts, leading to record low or high temperatures. This regional climate variability does not disprove long-term climate change. While 2008 was slightly cooler than 2007, partially due to a La Niña event, it was nonetheless the 10th-warmest year on record.

So what is going to happen now?

Will the *Washington Post* publish another letter from the WMO Secretary General debunking Will's anti-scientific claim and its misuse of WMO data? And then let Will do it again? Then publish another WMO letter?

Is the *Post* in the business of trying to inform its readers or does it just publish anything anybody writes? Does the editorial staff of the *Post* exercise any editorial judgment whatsoever?

Finally, most of Will's piece is an attack on compact fluorescent light bulbs. You won't be surprised to learn that it is as misinformed as everything else Will writes on this subject. Yes, as is the case with of all new products rushed to market—in this case, very low cost compact fluorescents—not all of them are high quality. They will get better, but they still save consumers big money and reduce pollution—and the tiny amount of mercury they contain locked away in their hardware is infinitely preferable to the vast quantities of mercury released into the air from coal consumption.

The bottom line about conservatives deniers like Will is that they have no plan at all to protect the health and well-being of your children and grandchildren and the next fifty generations from catastrophic global warming impacts. They just have lies and disdain for all things green and efficient.

The bottom line about the *Post* is that it would appear to have no journalistic standards at all for what it publishes on its editorial page and its letters page. Let me end with what Hilzoy of the *Washington Monthly* wrote about Will's first (!) piece:

> Where I come from, when someone writes something of the form: "P is not evidence for Q, and here's why," it is dishonest to quote that person saying P *and use that quote as evidence for* Q. If one of my students did this, I would grade her down considerably, and would drag her into my office for an unpleasant talk about basic scholarly standards. If she misused quotes in this way repeatedly, I might flunk her.

The *Washington Post* editorial staff has flunked Journalism 101. Fred Hiatt should be replaced.

CHAPTER 8

Diagnosing Someone With Anti-Scientific Syndrome (ASS)

Those who deny the climate science that makes clear global warming is an urgent problem or who seek to delay strong action to reduce greenhouse gas emissions have been more persuasive than climate scientists. These disinformers and delayers tend to be conservatives, so they have naturally been persuasive to conservatives.

A *National Journal* poll in June 2008 found that only 26 percent of GOP Congress members believe "it's been proven beyond a reasonable doubt that the Earth is warming because of man-made pollution." That matches their constituents—only 27 percent of Republicans say the Earth is warming because of human activity. Needless to say, if you don't believe humans are the cause of global warming, you're not going to believe that humans are the solution to global warming.

In this chapter, I'll look closer at the anti-scientific disinformers and what the polling tells us about their efforts. I will try to reserve the term "deniers" for the professional anti-scientific disinformers.

Diagnosing a Victim of Anti-Science Syndrome (ASS)

January 5, 2009

In this post I'm going to present the general diagnosis for "anti-science syndrome" (ASS). Like most syndromes, ASS is a collection of symptoms that individually may not be serious, but taken together can be quite dangerous—at least it can be dangerous to the health and well-being of humanity if enough people actually believe the victims.

One telltale symptom of ASS is that a website or a writer focuses their climate attacks on nonscientists. If that nonscientist is Al Gore, this symptom alone may be definitive.

The other key symptoms involve the repetition of long-debunked denier talking points, commonly without links to supporting material. Such repetition, which can border on the pathological, is a clear warning sign.

Scientists who kept restating and republishing things that had been widely debunked in the scientific literature for many, many years would quickly be diagnosed with ASS. Such people on the web are apparently heroes—at least to the right wing and/or easily duped.

If you suspect someone of ASS, look for the repeated use of the following phrases:

- Sunspots
- Hoax
- Medieval Warm Period
- Hockey Stick
- Michael Mann
- The climate is always changing
- Alarmist
- Temperature rises precede rises in carbon dioxide
- Pacific Decadal Oscillation

- Water vapor
- Cosmic rays
- Danish physicist Henrik Svensmark
- Ice Age was predicted in the 1970s
- Global cooling

Individually, some of these words and phrases are quite useful and indeed are commonly used by both scientists and nonscientists who are not anti-science. But the use of more than half of these in a single speech or article is pretty much a definitive diagnosis of ASS.

A newly prominent victim of ASS is Harold Ambler, who managed to get an anti-scientific climate denial article published on HuffingtonPost over the weekend: "Mr. Gore: Apology Accepted."

I was not originally planning to post on this (unsourced) collection of long-debunked denier talking points since, as regular readers know, my policy is not to waste time on the umpteenth debunking. Anyone who might be persuaded by Ambler's tripe can do a simple search for each myth on RealClimate.org or on this blog. Relevant ClimateProgress.org posts, to name but a few, include:

- Study: Sun's Contribution to Recent Warming Is "Negligible"
- Scientist: "Our Conclusions Were Misinterpreted" by Inhofe, CO_2— But Not the Sun—"Is Significantly Correlated" with Temperature since 1850
- UK Ministry of Defence: Global Warming Goes on, Deniers Are Deluded
- Study: Water-Vapor Feedback Is "Strong and Positive," So We Face "Warming of Several Degrees Celsius"
- Sorry Deniers, Hockey Stick Gets Longer, Stronger: Earth Hotter Now Than in Past 2,000 Years

For more detailed debunking (with links and citations) of every single myth that Ambler raises (without bothering to present links and citations) go to SkepticalScience.com.

As anti-scientific disinformers go, Ambler is quite lame. Separate from his long list of long-debunked denier talking points, who could possibly take seriously somebody who wrote the following:

> Mr. Gore has stated, regarding climate change, that "the science is in." Well, he is absolutely right about that, except for one tiny thing. It is the biggest whopper ever sold to the public in the history of humankind.

Such a statement is anti-scientific and anti-science in the most extreme sense. It accuses the scientific community broadly defined of deliberate fraud—and not just the community of climate scientists, but the leading National Academies of Science around the world (including ours); the American Geophysical Union, an organization of geophysicists that consists of more than 45,000 members; the American Meteorological Association; and the American Association for the Advancement of Science.

Such a statement accuses all of the member governments of the IPCC, including ours, of participating in that fraud, since they all sign off on the Assessment Reports word for word (see "Absolute MUST Read IPCC Report: Debate Over, Further Delay Fatal, Action Not Costly," in chapter 1). And, of course, Ambler's statement accuses all of the leading scientific journals of being in on this fraud, since the IPCC reports are primarily a review and synthesis of the published scientific literature.

So who could possibly take seriously Mr. Ambler? None other than "Best Science Blog" finalist WattsUpWithThat? Yes, Anthony Watts reprints Ambler's entire post—and does so approvingly:

> Shocker: Huffington Post Carries Climate Realist Essay
>
> Congratulations to Harold Ambler, who frequents here in comments, for breaking the climate "glass ceiling" at HuffPo. This essay is something I thought I'd never see there. Next stop: Daily Kos? —Anthony

To reprint Ambler's post and call it a "climate realist essay" makes Watts as anti-scientific as Ambler himself.

Scientist: "Our Conclusions Were Misinterpreted" by Inhofe, CO_2—But Not the Sun—"Is Significantly Correlated" with Temperature since 1850

December 12, 2008

The lead author of a new study says Inhofe's office mischaracterized her work with its blaring headline, "Study: Half of Warming Due to Sun!" Far from supporting Inhofe's denialist fantasies, the research, led by Anja Eichler, senior scientist at Switzerland's Paul Scherrer Institute, is actually one more piece of observation-driven analysis that strongly backs the reality of human-caused warming.

I pointed Eichler to the Senate website where an Inhofe staffer not only misstated her results but also concluded:

> Even if you try to stretch these numbers a little bit—but not unrealistically—you have to become sure that the participants of the Poznan [global climate negotiations] conference are lunatics.

Yes, on the basis of misrepresenting the work of one study, Inhofe's office calls the world's leading climate delegates "lunatics." But the study showed the exact opposite of what Inhofe's office said—and the climate delegates are working to stop humanity's self-destruction, while Inhofe are trying to accelerate it. So who are the crazy ones here?

Eichler replied to my e-mail:

> Thank you for informing us about the controversial discussion of our paper in your country. You are totally right that our conclusions were misinterpreted and we are a bit concerned about that.

I also posed her a couple of clarifying questions:

ROMM: Am I correct that your study was *not* saying human-caused emissions were *not* the major factor driving the temperature record in the past century?

EICHLER: Yes, this is correct. We did a strong differentiation between preindustrial (1250–1850) time and the last 150 years. In the preindustrial time we found a strong correlation between the solar activity proxy and our temperature, suggesting solar forcing as a main force for temperature change in this time. However, the correlation between the solar activity proxy and Altai temperature is *not* significant anymore for the last 150 years. In this time the increase in the CO_2 concentrations is significantly correlated with our temperature.

ROMM: Am I correct that your final sentence [in the paper] was merely saying that your results suggest the Sun was responsible for under 50 percent of the warming since 1900, but you were *not* saying your results shows that the Sun was in fact responsible for half the warming.

EICHLER: This is also absolutely correct.

She added that "uncertainties of our data" do not allow it to be used to give an exact percentage for how much solar activity was responsible for the warming in the past century. A 2008 study concluded that the Sun's contribution to recent warming is "negligible." And a 2007 study concluded:

> Here we show that over the past twenty years, all the trends in the Sun that could have had an influence on the Earth's climate have been in the opposite direction to that required to explain the observed rise in global mean temperatures.

And a 2006 study found "The variations measured from spacecraft since 1978 are too small to have contributed appreciably to accelerated global warming over the past thirty years." And a 2005 study concluded:

> During these last thirty years the solar total irradiance, solar UV irradiance and cosmic ray flux has not shown any significant secular trend, so that at least this most recent warming episode must have another source.

The fact that Inhofe's office would completely misstate the results of the study is nothing new or terribly interesting. But the conclusions of the study are quite intriguing in that they underscore the key point that the deniers refuse to accept: The Earth's temperature does not change randomly—it changes when it is driven to do so by an external forcing.

Yes, deniers—some of whom comment on this website—the Earth has had brief warming and cooling periods since 1250. But those temperature changes were *not* random. They were largely responses to changes in the solar radiation hitting the Earth (which is itself affected by volcanoes).

Now human-caused emissions are driving climate change to dangerous levels with forcings that dwarf previous natural forcings both in speed and scale. Humans have boosted CO_2 concentrations in the past two centuries 14,000 times faster than nature changed those levels in the last 600,000 years, which has "put the [climate] system entirely out of equilibrium." And that's why the time to act is now.

The Deniers Are Winning, Especially with the GOP

May 9, 2008

The science is clear about the reality of global warming and the fact that humans are the dominant cause. But, sadly, that isn't clear to most Republicans.

Anybody who thinks the public debate is over—anybody who thinks anti-scientific disinformation doesn't work—should look at the latest poll results from the Pew Research Center:

> The proportion of Americans who say that the Earth is getting warmer has decreased modestly since January 2007, mostly because of a decline among Republicans.

Fewer Republicans Say the Earth Is Warming

% saying there is solid evidence of global warming:	Jan 2007 %	April 2008 %	*Change*	2008 N
Total	77	71	-6	1502
Republicans	62	49	-13	412
Democrats	86	84	-2	521
Independents	78	75	-3	462

Only 49 percent of Republican now even believe that the Earth is warming! Thank you so much deniers, delayers, and mainstream media.

Even more worrisome is just how many people don't believe humans are the cause of warming:

> Roughly half of Americans (47 percent) say the earth is warming because of human activity, such as the burning of fossil fuels [and only 27 percent of Republicans]. But nearly as many people (45 percent) say that rising global temperatures are either mostly caused by natural

> environmental patterns (18 percent), say they do not know the cause of warming (6 percent), or say that no solid evidence of warming exists (21 percent).

I'd like to thank the media, especially NBC news, for contributing to this core talking point of the disinformers (As Dateline NBC put it, "Whatever the Cause . . . Global Warming Is a Reality").

The science makes unequivocally clear that the health and well-being of billions of people (and most species) are at grave risk from continued unrestricted human emissions of greenhouse gases.

But who could possibly believe that there are so many credible-sounding people, including major public leaders in the conservative movement, who would so strongly argue that

1) The Earth is not warming, and/or
2) Humans are not a major cause of whatever warming is occurring, and/or
3) The problem is not an urgent one because the impacts are distant and tolerable, and/or
4) The solution is painful if not impossible with existing technologies anyway, and/or
5) Adaptation is a better strategy than mitigation.

It is hard to believe—indeed it is almost impossible to believe—that so many seemingly serious people could say those things.

And it has proven almost impossible for the traditional media to deal with, as we've seen.

I don't have any easy answers to offer in this post. Shaming the traditional media doesn't seem to work because they are mostly shameless—indeed the vast majority of journalists wear it as a badge of honor that they are criticized equally by "both sides."

I suppose the only answer is vigilance. The cost of losing is simply too high.

The Deniers Are Winning, but Only with the GOP

September 18, 2008

Turns out you can fool some of the people all of the time—if those people are conservatives.

I have previously argued "the deniers are winning, *especially* with the GOP," which received more than 500 comments, my most ever. Now *Environment* magazine has published an analysis in their September/October 2008 issue that suggests the deniers are winning *only* with the GOP. This analysis should be especially alarming to scientists.

Is Global Warming Occurring?

As the figure shows, despite the steadily growing observational evidence that global warming has begun—indeed that it is occurring faster than expected—conservatives have actually been less than oblivious. The global cooling lie has worked—with the GOP.*

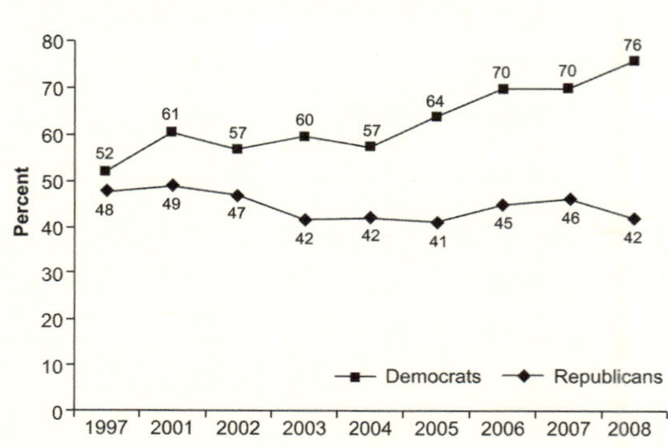

* All of the figures here come from Gallup polling over the past decade.

That shouldn't be too surprising, I guess, since the disinformation campaign aimed at blocking climate action comes primarily from the conservative movement itself:

> A significant part of the U.S. conservative movement—made up of conservative foundations, think tanks, media, and public intellectuals—mobilized in the 1990s to challenge both climate science and climate policy.

Needless to say, those denial sources have more credibility with conservatives, so it is only natural that so many conservatives have been duped. Indeed, some conservatives simply adopt the positions of conservative intellectuals without doing any thinking of their own.

Is There a Scientific Consensus?

A second figure shows another remarkable partisan divide.

> Republican spokespersons and conservative commentators continue to challenge the scientific consensus on global warming by highlighting the views of a modest number of skeptic or "contrarian" scientists who question the IPCC's conclusions. One result is that in their efforts to provide "balanced coverage, U.S. media have given disproportionate attention to these skeptics, creating the impression of less scientific consensus on global warming than exists within the mainstream scientific community."†

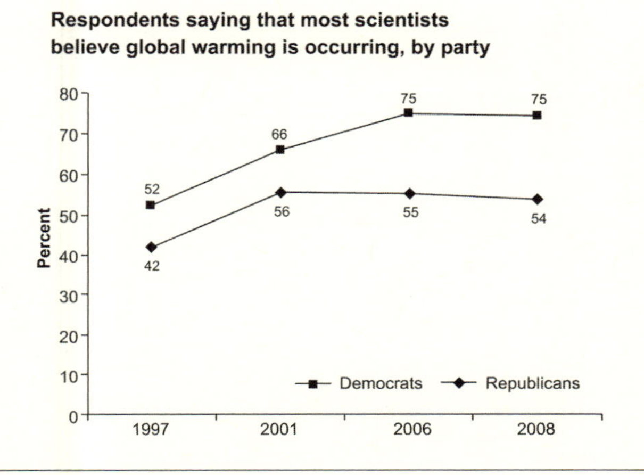

† I don't like the term "consensus," as I've written. I prefer "understanding."

To see just how remarkable those data are, compare the two figures. In 1997, some 52 percent of Dems said the effects of global warming have already begun *and* 52 percent said most scientists believe global warming is occurring. In 2008, now 76 percent say warming has begun *and* 75 percent say most scientists believe warming is occurring. Makes sense. Dems believe most scientists.

As for Republicans, in 1997 some 42 percent said warming had begun and 48 percent said most scientists believe warming is occurring—a modest six-point differential. By 2008, the percentage of Republicans saying the effects of global warming have already begun had *dropped* to a mere 42 percent (an amazing stat in its own right given the painfully obvious evidence to the contrary). But the percentage saying most scientists believe global warming is occurring had *risen* to 54 percent—a stunning twelve-point differential.

In short, a significant and growing number of Republicans—one in eight as of 2008—simply don't believe what they know most scientists believe. That is quite alarming news, given that it is inconceivable the nation will take the very strong action needed to avert catastrophe unless it comes to believe what most scientists believe, namely that we are in big, big trouble and can delay no further.

The time for scientists to deal with this abject failure to communicate is yesterday!

What about the partisan divide over whether humans are to blame for most of the warming that has occurred? These results may be the most depressing of all:

Human-Caused or Natural Change?

> In 2008, 58 percent of the surveyed population sees global warming as due more to human activities than natural causes, slightly lower than the 61 percent giving this response in 2001. . . . This near-stability in the overall population hides differing trends among Republicans and Democrats.

A large majority of Democrats has understood for a decade that humans are the dominant cause of global warming. Only two in five Republicans now believe this, even though it is a central conclusion of the recent, definitive 2007 IPCC report.

This is really the core scientific and political issue: If you believe the sharp *increases* in human-caused emissions of greenhouse gases during the last century are not the primary cause of global warming then it would seem obvious you will believe that sharp *decreases* in human-caused emissions are not the solution. And this is a core reason why congressional conservatives seem increasingly unwilling to back serious climate action. Unless these numbers change substantially, climate action will remain a very polarized issue politically.

Is Media Coverage Exaggerated?

The main sources of denial come from conservative sources outside of the traditional, mainstream media. So it is only natural that Republicans would be increasingly skeptical of the mainstream media, whose coverage, though still quite lame, as we've seen, is at least somewhat driven by the rapidly growing quantity and quality of observational evidence along with the increasingly strong scientific understanding of climate science. In the past decade, the relatively low percentage of Democrats saying "the seriousness of global warming is generally exaggerated in the news" has stayed flat, whereas it has risen steadily from 37 percent to 59 percent among Republicans.

And so we have what can be described only as an increasingly grim political situation. Most Republicans don't believe the effects of global warming are already here, and they don't believe humans are the primary cause of temperature changes in the last century.

The two obvious sources of information that might change these dangerously mistaken views are the scientific community and the traditional news. But Republicans have become increasingly skeptical of both those sources over time.

That really leaves only one source of "information" that might change the views of Republicans and that is the leadership of the conservative movement itself—conservative politicians, conservative think tanks, conservative media, and conservative pundits. Until they not only reverse their position completely but also actively spread this reversed position to the faithful, this country will find it almost impossible to adopt the very strong government-led policies needed to avert humanity's self-destruction.

Yet this presidential campaign suggests the tiny portion of the conservative movement that actually seemed to understand global warming is now moving in the wrong direction.

I don't have any easy answers to offer in this post. I do believe that if the conservative movement continues to strongly oppose serious climate action, then it will ultimately be destroyed by that self-destructive myopia. But that will be very small comfort to the billions and billions of people whose lives are ruined by catastrophic global warming in the coming decades and centuries.

Irony-Gate: Viscount Monckton, a British Peer, Says His Paper Was Peer-Reviewed by a Scientist

July 20, 2008

"The Viscount Monckton of Brenchley" is full of himself. Before casting a wary eye on his new ribaldry, however, let me note that he has been repeatedly debunked by actual scientists.

But I digress. The Viscount Monckton of Brenchley, as he prefers to call himself, or TVMOB, as I will call him because, damn, the acronym is just too sweet, has penned an epistle to the president of the American Physical Society (which you can peruse on my blog).

TVMOB is displeased with the new APS disclaimer on his article: "The following article has not undergone any scientific peer review. Its conclusions are in disagreement with the overwhelming opinion of the world scientific community. The Council of the American Physical Society disagrees with this article's conclusions."

TVMOB writes, "This seems discourteous." You see, TVMOB holds the view that peer review occurs if his article gets suggested edits by a coeditor who happens to be a scientist.

Let me not make the obvious point that being edited by an editor ain't scientific peer review. You can read the editor's requested edits on page 2 of TVMOB's letter. Anybody who has actually been peer-reviewed will note that the proposed edits aren't anything close to what a peer-reviewed set of comments looks like, especially for an analysis as flawed as this one.

Since TVMOB's letter is straight out of Monty Python, let me rather make the point in kind that a peer is "a person who holds any of the five grades of the British nobility: duke, marquess, earl, viscount, and baron."

By that definition, I am sure that TVMOB's paper was not given proper peer review. Indeed, I'm not certain TVMOB has a proper peer on this Earth. Perhaps Senator Inhofe or President Bush.

But pity the poor modern British viscount who whines in his letter, "I had expended considerable labor, without having been offered or having requested any honorarium." Join the club, buddy. Since when do you think scientific newsletters pay you a nickel? Oh, I forgot. You aren't a scientist.

I especially love the conclusion to his epistle:

> Please either remove the offending red-flag text at once or let me have the name and qualifications of the member of the Council or advisor to it who considered my paper before the Council ordered the offending text to be posted above my paper; a copy of this rapporteur's findings and ratio decidendi; the date of the Council meeting at which the findings were presented; a copy of the minutes of the discussion; and a copy of the text of the Council's decision, together with the names of those present at the meeting. If the Council has not scientifically evaluated or formally considered my paper, may I ask with what credible scientific justification, and on whose authority, the offending text asserts primo, that the paper had not been scientifically reviewed when it had; secundo, that its conclusions disagree with what is said (on no evidence) to be the "overwhelming opinion of the world scientific community"; and, tertio, that "The Council of the American Physical Society disagrees with this article's conclusions"? Which of my conclusions does the Council disagree with, and on what scientific grounds (if any)?
>
> Having regard to the circumstances, surely the Council owes me an apology?
>
> Yours truly, THE VISCOUNT MONCKTON OF BRENCHLEY

Pistols at dawn, perhaps?

One denier website actually labels this "PeerGate scandal." But I believe they have missed the irony.

Should you be interested in learning more about TVMOB, go to the Science & Public Policy Institute website where he is chief policy adviser. You will learn he has astonishing scientific credentials such as a "Nobel prize pin," because he commented on the IPCC Fourth assessment report. This has "earned him the status of Nobel Peace Laureate. His Nobel prize pin, made of gold recovered from a physics experiment, was presented to him by the Emeritus Professor of Physics at the University of Rochester, New York." Also "his limpid analysis of the climate-feedback factor was published on the famous climate blog of Roger Pielke, Sr." I kid you not.

Monty Python is alive and well. Oh, and TVMOB knows how to use the words "primo" and "secundo" and "tertio." Some of us can only dream of such scientific achievements.

Finally, if his writing has made you a fan of TVMOB, you can go to HouseOfNames.com and purchase products labeled with the Monckton family crest, including mouse pads.

CHAPTER 9

Why Are Progressives So Lousy at Messaging?

In the last decade, scientists have apparently become less convincing to Republicans than the anti-scientific disinformers have been. As the polling discussed in the last chapter reveals, a significant and growing number of Republicans—one in eight as of 2008—simply don't believe what they know most scientists believe. The disinformers apparently have gotten more effective at messaging while scientists, progressives, and environmentalists have gotten worse.

In part, this has occurred because there is an organized disinformation campaign promoted by conservative think tanks like the Competitive Enterprise Institute and well-funded by fossil fuel companies like ExxonMobil, with key messages broadcast repeatedly by conservative pundits and politicians like George Will and Rush Limbaugh and Senator James Inhofe (R-OIL).

At the same time, the status quo media has treated this more as a political issue than a scientific one, thereby necessitating in their view a "balanced" presentation of both sides, notwithstanding the fact that the overwhelming majority of scientists understand humans are warming the planet and dangerously so. Also, increasingly profit-driven media have been abdicating their role in science education. Science writers Chris Mooney and scientist Sheril Kirshenbaum offer these grim statistics in their recent book, *Unscientific America: How Scientific Illiteracy Threatens our Future*:

- For every five hours of cable news, one minute is devoted to science.
- Forty-six percent of Americans believe the Earth is less than 10,000 years old.
- The number of newspapers with science sections has shrunken by two-thirds in the last twenty years.
- Just 18 percent of Americans know a scientist personally.
- The overwhelming majority of Americans polled in late 2007 either couldn't name a scientific role model or named "people who are either not scientists or not alive."

As the media is abandoning science coverage, more and more scientists are starting to speak out, as discussed in chapter 2. But progressives in general and scientists in particular aren't great at messaging. And in part that's because well-meaning but misguided pollsters have convinced some progressives and environmentalists they should downplay talk about global warming and its impacts, which amounts to unilateral disarmament in the battle to explain to the public what will happen if we continue on our current path of unrestricted greenhouse gas emissions.

Why Scientists Aren't More Persuasive, Part 1

September 30, 2008

> Of all the talents bestowed upon men, none is so precious as the gift of oratory. He who enjoys it wields a power more durable than that of a great king. . . . The subtle art of combining the various elements that separately mean nothing and collectively mean so much in an harmonious proportion is known to very few. . . . [T]he student of rhetoric may indulge the hope that Nature will finally yield to observation and perseverance, the key to the hearts of men.

So wrote a twenty-three-year-old Winston Churchill in a brilliant, unpublished essay, "The Scaffolding of Rhetoric."

The ever-worsening reality of human-caused global warming is driving more and more scientists to become desperate about our future. Yet poll after poll shows that scientists and those who accept scientific understanding as the basis for action on climate change are failing to persuade large segments of society about the urgent need to act.

Anyone who wants to understand—and change—the politics of global warming must understand why the disinformers, delayers, and inactivists are so persuasive in the public debate, and why scientists and scientific-minded people are not. A key part of the answer, I believe, is that while science and logic are powerful systematic tools for understanding the world, they are no match in the public realm for the twenty-five-century-old art of verbal persuasion: rhetoric.

Logic might be described as the art of influencing minds with the facts, whereas rhetoric is the art of influencing both the hearts and minds of listeners with the figures of speech. The figures are the catalog of the different, effective ways that we talk—they include alliteration and other forms of repetition, metaphor, irony, and the like. The goal is to sound believable. As Aristotle wrote in *Rhetoric*, "aptness of language is one thing that makes people believe in the truth of your story."

The rhetorical figures have been widely studied by marketers and social scientists. They turn out to "constitute basic schemes by which people conceptualize their experience and the external world," as one psychologist put it. We think in figures, and so the figures can be used to change the way we think. That's why political speech writers use them. To help level the rhetorical playing field in the global warming debate, I will highlight the three rhetorical elements that are essential to modern political persuasion.

First: simple language. Contrary to popular misconception, rhetoric is not big words; it's small words. Churchill understood this at the age of twenty-three:

> The unreflecting often imagine that the effects of oratory are produced by the use of long words. . . . The shorter words of a language are usually the more ancient. Their meaning is more ingrained in the national character and they appeal with greater force to simple understandings than words recently introduced from the Latin and the Greek. All the speeches of great English rhetoricians . . . display an uniform preference for short, homely words of common usage. . . .

We hear the truth of his advice in the words that linger with us from all of the great speeches: "Judge not that ye be not judged," "To be or not to be," "lend me your ears," "Four score and seven years ago," "blood, toil, tears, and sweat," "I have a dream."

In short, simple words and simple slogans work.

Second, repetition, repetition, repetition. Repetition makes words and phrases stick in the mind. Repetition is so important to rhetoric that there are four dozen figures of speech describing different kinds of repetition. The most elemental figure of repetition is *alliteration* (from the Latin for "repeating the same letter"), as in "compassionate conservative." Repetition, or "staying on message," in modern political parlance, remains the essential rhetorical strategy. As Frank Luntz—the bane of climate progressives, but an undeniably astute conservative messaging guru—has said:

> There's a simple rule: You say it again, and you say it again, and you say it again, and you say it again, and you say it again, and then again and again and again and again, and about the time that you're absolutely sick of saying it is about the time that your target audience has heard it for the first time.

Third, the skillful use of *tropes* (from the Greek for *turn*), figures that change or turn the meaning of a word away from its literal meaning. The two most important tropes, I believe, are metaphor and irony.

"To be a master of metaphor," Aristotle writes in *Poetics*, is "a sign of genius, since a good metaphor implies intuitive perception of the similarity in dissimilars." When Bush said in 2006 that the nation was "addicted to oil," he was speaking metaphorically. Curing an addiction, however, requires far stronger medicine than the president proposed.

Science, Climate, and Rhetoric

Rhetoric works, and it works because it is systematic. As Churchill wrote, "The subtle art of combining the various elements that separately mean nothing and collectively mean so much in an harmonious proportion is known to very few."

Unfortunately, the major player in the climate debate, the scientific community, is not good at persuasive speech. Scientists might even be described as anti-rhetoricians since they avoid all of its key elements.

Few scientists are known for simple language. As the physicist Mark Bowen writes in *Thin Ice*, his book about glaciologist Lonnie Thompson:

> Scientists have an annoying habit of backing off when they're asked to make a plain statement, and climatologists tend to be worse than most.

Most scientists do not like to repeat themselves because it implies that they aren't sure of what they are saying. Scientists like to focus on the things that they don't know, since that is the cutting edge of scientific research. So they don't keep repeating the things that they do know, which is one reason the public and the media often don't hear from scientists about the strong areas of agreement on global warming.

The disinformers are so good at repetition that they continue to repeat myths long after they have been debunked by scientists. Scientists, and the media, grow weary of repeatedly debunking the same lies, the same nonsensical myths. But that, of course, only encourages the disinformers to keep repeating those myths.

Like my two-year-old daughter, they know that if they just keep repeating the same thing over and over and over and over again, they will eventually get their way. And they have. Of course, when your "way" is just to get people to keep doing the same thing they have been doing for decades (i.e., nothing), your messaging task is considerably easier because the default position of most people, the media, and policy makers is "do nothing."

Finally, scientific training, at least as I experienced it, emphasizes sticking to facts and speaking literally, as opposed to figuratively or metaphorically. Scientific debates are won by those whose theory best explains the facts, not by those who are the most gifted speakers. This view of science is perhaps best summed up in the motto of the Royal Society of London, one of the world's oldest scientific academies (founded in 1660), *Nullius in verba*: take nobody's word. Words alone are not science.

Scientists who are also great public communicators, like Carl Sagan or Richard Feynman, have grown scarcer as science has become increasingly specialized. Moreover, the media likes the glib and the dramatic, which is the style most scientists deliberately avoid. As Jared Diamond (author of *Collapse*) wrote in a must-read 1997 article on scientific messaging (or the lack thereof), "Scientists who do communicate effectively with the public often find their colleagues responding with scorn, and even punishing them in ways that affect their careers." After Carl Sagan became famous, he was rejected for membership in the National Academy of Sciences in a special vote. This became widely known, and, Diamond writes, "Every scientist is capable of recognizing the obvious implications for his or her self-interest."

Scientists who have been outspoken about global warming have been repeatedly attacked as having a "political agenda." As one 2006 journal article explained, "For a scientist whose reputation is largely invested in peer-reviewed publications and the citations thereof, there is little professional payoff for getting involved in debates that mix science and politics."

Not surprisingly, many climate scientists shy away from the public debate. At the same time, the Bush administration has muzzled many climate scientists working for the U.S. government. As a result, science journalists, not practicing scientists, are almost always the ones explaining global warming to the public. Unfortunately, the media is cutting back on science reporting in general, and finds reporting climate science particularly problematic.

It is not remarkable, then, that the American public is so uninformed about global warming, so vulnerable to what might be called the conservative crusade against climate.

Why Scientists Aren't More Persuasive, Part 2: Why Deniers Out-Debate "Smart Talkers"

October 13, 2008

In 2007, NPR broadcast a now-infamous climate debate on the proposition "Global warming is not a crisis." In theory, this sounds like an easy win for the "nay" side—"crisis" is obviously the mildest of words to describe the greatest preventable existential threat to the health and well-being of future generations.

But in practice such debates are almost unwinnable, even by those who are good at debating in public, a group that does not include very many scientists. As noted in Part 1, scientists are lousy at rhetoric, the art of persuasion. Significantly, rhetoric was discovered and developed by the Greeks and Romans in part to help them win debates, so it follows that modern debates are also won by those who are better at using the strategies and tactics of rhetoric. In his dialogue *Gorgias* about the master rhetorician, Plato gives him a speech that dramatizes the awesome power of rhetoric:

> If a rhetorician and a doctor visited any city you like to name and they had to contend in argument before the Assembly or any other gathering as to which of the two should be chosen as doctor, the doctor would be nowhere, but the man who could speak would be chosen, if he so wished.

So a rhetorician could persuade any audience, no matter how intelligent, that he or she was more of a doctor than a real doctor. No surprise, then, that someone skilled in rhetoric can beat a scientist in a debate on climate.

The 2007 debate had, "speaking for the motion: Michael Crichton, Richard S. Lindzen, Philip Stott" and "speaking against the motion: Brenda Ekwurzel, Gavin Schmidt, Richard C. J. Somerville." The painfully inevitable result as announced by NPR's Brian Lehrer at the end:

> And now the results of our debate. After our debaters did their best to

> sway you . . . you went from, 30 percent for the motion that global warming is not a crisis, from 30 percent to 46 percent. [APPLAUSE] Against the motion, went from 57 percent to 42 percent . . . [SCATTERED APPLAUSE].

A few more debates like that and we can all buy beachfront property in Baton Rouge.

Personally, I still do one-on-one debates from time to time, although they are almost unwinnable against a sophisticated disinformer or delayer, like, say Bjørn Lomborg. But a three-on-three is quite counterproductive, since the other side will just go after your weak link(s). The other flaw in this debate is the proposition. "Crisis" is a losing word—sorry Al Gore—a word the public has grown tired of, since it's been applied to too many (every?) major public policy problem in the last two decades.

In this post, I'll talk a little bit more about why "smart-talkers" like scientists don't tend to win debates. I won't critique the climate scientists in the 2007 debate, but comment instead on two of the disinformers/delayers. Stott spends a considerable amount of time pushing the favorite disinformer narrative that just a few decades ago, scientists believed the climate was cooling but now they believe it's warming. I will explain below why someone like Stott who has spent ten years using "modern techniques to deconstruct grand environmental narratives, like global warming," would devote so much time to repeating such a long-debunked myth.

Even more fascinating is the opening statement from the one nonscientist in the debate, the late Michael Crichton, who obviously became very rich precisely because he knows how to put together (fictional) narratives that are compelling to millions of people. He adopts the classic everyman position that is classic old-school rhetoric:

> I myself, uh, just a few years ago, held the kinds of views that I, uh, expect most of you in this room hold. That's to say, I had a very conventional view about the environment. I thought it was going to hell. I thought human beings were responsible and I thought we had to do something about it. I hadn't actually looked at any environmental issues in detail but I have that general view. And so in 2000, when I read an

> article that suggested that the evidence for global warming might not be quite as firm as people said, I immediately dismissed it. Not believe in global warming? That's ridiculous. How could you have such an idea? Are you going to try and tell me that the planet isn't getting warmer? I know it's getting warmer. . . . I spent thirty years in California. We used to have something called June gloom. Now it's more like May, June, July, August gloom with September, October, November gloom added in. The weather is very different.
>
> However, because I look for trouble, um, I went at a certain point and started looking at the temperature records. And I was very surprised at what I found. The first thing that I discovered, which Dick has already told you, is that the increase in temperatures so far over the last hundred years, is on the order of six-tenths of a degree Celsius, about a degree Fahrenheit. I hadn't really thought, when we talked about global warming, about how much global warming really was taking place. . . .

Bullshit? Yes. Persuasive? I'm afraid so. Crichton is identifying himself with the audience—he once believed like they do, but then, gosh darn it, he went looking for trouble and found the actual data. This rhetorical strategy, and Stott's, is not just decades old, not just centuries old, it is literally millennia old.

Let me bring presidential politics into this because, frankly, that is the origin of much of my analysis. Scientists and progressives and Democratic politicians have historically lost debates because they made two fundamental mistakes: First, they have treated the debates as if they were high school or college debates, which are won primarily on the merits of the arguments and volume of evidence presented.

Second, relatedly, they seem to think that appearing smarter than your opponent is a *winning* strategy, whereas conservatives understand and have repeatedly demonstrated it is a *losing* strategy. This fact was very well understood by the masters of persuasive language from ancient Greece and Rome through Elizabethans like Shakespeare and by skilled debaters like Lincoln and Churchill, as we will see.

Debates are typically won by the candidate who presents the most compelling and persuasive character. If I can convince you I'm an honest, straight talker, you'll believe what else I say. If I can't, you won't.

Debates are not usually won on factual or policy merits, in part because listeners aren't in a position to adjudicate sometimes subtle differences between complex positions—what exactly was the difference between Clinton's health care plan and Obama's? And what exactly is the difference between carbon dioxide emissions and carbon dioxide concentrations?—and because those who are undecided on an issue are typically skeptical of all advocates, especially self-styled "experts." They assume everybody exaggerates to defend their position. In any case, if I don't convince you I'm honest, my stated positions can't possibly matter.

The rest of this post will explain why (those who appear to be) straight talkers beat smart talkers every time, ending with a discussion of the 2004 election.

A History of Faking Straight Talk

A core strategy of rhetoric is to avoid seeming like a smarty-pants, to avoid appearing like Carter, Dukakis, Gore, and Kerry—a highly educated (i.e., elite), wonkish speaker—but rather to appear a plainspoken man of the people.

Shakespeare—a master of rhetoric who knew more than 200 figures of speech, like all middle-class Elizabethans—understood that very well. That's why he has Mark Antony say in one of the great debate speeches of all time, his famous "Friends, Romans, countrymen" response to Brutus in the Roman Forum: "I am no orator, as Brutus is, But—as you know me all—a plain blunt man."

Is it coincidental that the only ones to use the word "rhetoric" in the 2004 presidential debates were George W. Bush and Dick Cheney? In the vice presidential debate, Cheney said to his Democratic rival, Senator John Edwards, "Your rhetoric, Senator, would be a lot more credible if there was a record to back it up." In the final debate, Bush twice repeated almost verbatim the same accusation about Kerry: "His rhetoric doesn't match his record," and again, "His record in the United States Senate does not match his rhetoric." This was only a small salvo in the Bush team's war on Kerry's language.

It is a mark of wily orators that they accuse their opponents of being rhetoricians. Winston Churchill, who wrote a treatise on the use of rhetoric in political speech at the age of twenty-three, himself once opened an attack on his political opponents, saying "These professional intellectuals who revel in decimals and polysyllables."

Returning to the Roman Forum, Marc Antony says:

For I have neither wit, nor words, nor worth,
Action, nor utterance, nor the power of speech,
To stir men's blood: I only speak right on;
I tell you that which you yourselves do know

So Antony is a man of the people, just reminding them of what they already know. Antony was, in fact, a patrician, like Bush. Indeed, Antony was a student of rhetoric, but his repeated use of one-syllable words lends credibility to his blunt sincerity. It is a mark of first-rate orators that they deny eloquence.

Lincoln was a "plain homespun" speaker, or so goes the legend, a legend he himself worked hard to create. In a December 1859 autobiographical sketch provided to a Pennsylvania newspaper, Lincoln explained how his father grew up "literally without education." Lincoln described growing up in "a wild region, with many bears and other wild animals still in the woods. . . . There were some schools, so called." He offers one especially colorful spin: "If a stranger supposed to understand Latin, happened to sojourn in the neighborhood, he was looked upon as a wizard." No fancy talkers here. Lincoln modestly explains the result of the little schooling he had: "Of course when I came of age, I did not know much." And after that, "I have not been to school since. The little advance I now have upon this store of education, I have picked up from time to time under the pressure of necessity." All this from a man who in the previous year had proven himself to be one of America's great orators in the Lincoln-Douglas debates and who during the course of his presidency would demonstrate the most sophisticated grasp of rhetoric of any U.S. president, before or since.

Lincoln opened his masterful February 1859 Cooper Union speech echoing Shakespeare's Antony: "The facts with which I shall deal this evening are mainly old and familiar; nor is there anything new in the general use I shall make of them." (In Antony's own words, "I only speak right on; I tell you that which you yourselves do know.") These are the words of a man who had memorized Shakespeare from William Scott's *Lessons in Elocution*, a treatise that included Antony's famous speech.

Does this sound, from Crichton, a little familiar?

> I myself, uh, just a few years ago, held the kinds of views that I, uh, expect most of you in this room hold.

If you want to switch people's viewpoints, pretend like you once held their views. It is a twofer. First, you can pretend you're just like one of them. Second, you draw people into the narrative, since they become intrigued about how someone who used to believe as they did now believes differently. Classic storytelling—you need to create a hook for the listener early on or they will tune out.

Returning to rhetoric, the master orator who denies eloquence was such a commonplace by the sixteenth century that Shakespeare resorted to it repeatedly. Consider his King Henry V, a master of oratory, who delivered the most famous pre-battle speech in the English-language:

> *We few, we happy few, we band of brothers;*
> *For he today that sheds his blood with me*
> *Shall be my brother*

After the British triumph at Agincourt, King Henry V woos Katherine, the daughter of the French king. Yet, even though Kate's hand was one of Henry's conditions for peace, the master of rhetoric still treats us to his tricks.

When Kate says she doesn't speak English well, Henry says he's glad, "for, if thou couldst, thou wouldst find me such a plain king that thou wouldst think I had sold my farm to buy my crown." He's just like a farmer, a man of the people. He adds, "But, before God, Kate, I cannot look greenly nor gasp out my eloquence, nor I have no cunning in protestation; only downright oaths, which I never use till urged, nor never break for urging." Like Antony, he disingenuously denies eloquence. The reason orators use this trick: Being blunt and ineloquent means they must be honest and steadfast.

Here is Bush in his Orlando campaign speech on October 30, 2004:

> Sometimes I'm a little too blunt—I get that from my mother. [Huge Cheers] Sometimes I mangle the English language—I get that from my dad. [Laughter and Cheers]. But you always know where I stand. You can't say that for my opponent.

For a blunt language-mangler, that's surprisingly old-school—very old school—rhetoric.

Henry urges Kate to "take a fellow of plain and uncoin'd constancy, for he perforce must do thee right, because he hath not the gift to woo in other places." Because he is not a clever orator, he must be an honest and constant man. Then Henry compares himself to an imaginary rival: "For these fellows of infinite tongue, that can rhyme themselves into ladies' favours, they do always reason themselves out again." In short, the other guys are flip-floppers and liars. They talk smarter than I do, but that's exactly why you can't trust them.

This is precisely why the disinformers like Stott and Crichton love to repeat the global cooling myth, love to say, as Crichton has one of his fictional environmentalists claim in *State of Fear*, "In the 1970s all the climate scientists believed an ice age was coming."

This clever and popular attack tries to make present global-warming fears seem faddish, saying current climate science is nothing more than finger-in-the-wind guessing. This attack appeals especially to conservatives who want to link their attack on climate scientists to their favorite attack against progressive presidential candidates—that they are flip-floppers. It's been debunked time and time again.*

Consider Bush's stump speech in Wilmington, Ohio, the day before the election, discussing his September 2003 request for $87 billion in Iraq war funding and Kerry's vote: "And then he entered the flip-flop Hall of Fame by

---

* A major 2008 literature review in the Bulletin of the American Meteorological Society concluded, "There was no scientific consensus in the 1970s that the Earth was headed into an imminent ice age. Indeed, the possibility of anthropogenic warming dominated the peer-reviewed literature even then."

saying this: 'I actually did vote for the $87 billion right before I voted against it.' I haven't spent a lot of time in the coffee shops around here, but I bet you a lot of people don't talk that way." In Burgettstown, two hours later he said, "I doubt many people in western Pennsylvania talk that way." In Sioux City, Iowa, a few hours later, "I haven't spent much time in the coffee shops around here, but I feel pretty comfortable in predicting that not many people talk like that in Sioux land." And in Albuquerque, he said, "I have spent a lot of time in New Mexico, and I've never heard a person talk that way."

Sarah Palin, in her stump speech, makes an almost identical criticism of Obama: "We tend to prefer candidates who don't talk about us one way in Scranton and another way in San Francisco." He is not one of us. He's two-faced. Yes, it may seem laughable coming from the Palin-McCain team, but even laughable works when it uses the tools of rhetoric—Palin here [or her speechwriter] is using *antithesis*—placing words or ideas in contrast or opposition, one of Lincoln's favorite rhetorical devices: "with malice toward none; with charity for all." And she is placing Obama into a very old narrative about liars, flip-floppers, and Democratic candidates for president.

Kerry's self-defining quote—"I actually did vote for the $87 billion right before I voted against it"—has the powerful elements of eloquence. Sadly for Kerry, this is the precise reason it stuck in the mind. It has the repetition and sound of two memorable figures found in famous political quotes, antithesis ("voted for" versus "voted against") and *chiasmus*, words repeated in inverse order (in this case, "I . . . vote for" and "before I voted"). Little wonder it was ripe for exploitation through repetition and sarcasm.

President Bush in 2004 had everything down cold that we expect from a master rhetorician: The repeated simple words, the repeated phrases, and the message that his opponent is inconsistent and inconstant because he's too clever by half and doesn't talk the way you and I do. Yet at the same time, Bush managed to leave the impression that he himself is rather slow and inarticulate. Ironically, the (all-too-many) Democrats who attacked Bush as being stupid merely gave him a free pass on all his lying and made him seem more genuine and credible to many voters.

As hard as it can be sometimes—and even I fall into the trap from time to time—it simply makes no sense whatsoever to attack your opponents as being stupid. Call them liars before calling them stupid.

Why did Kerry flip flop? Bush had a simple answer. The president told every audience that Kerry's most revealing explanation "was when he said, the whole thing was a complicated matter. My fellow Americans, there is nothing complicated about supporting our troops in combat."

Rhetoric retains the power to move real people. In a 2005 post-election analysis, journalism professor Danner quotes one Dr. Richardson-Pinto saying to him at Bush's Orlando rally: "It doesn't matter if the man [Kerry] can talk. Sometimes, when someone's real articulate, you can't trust what he says, you know?" And Richardson-Pinto is a doctor, someone whose credibility depends on being articulate.

So, yes, being smart, talking smart, and using big words may impress some in the audience—but most likely only those who already agree with you. It may cost you credibility with the very people you are trying to reach.

I fully understand that many scientists don't want to spend the time needed to learn how to be persuasive to nonscientists. Indeed, Part 1 discusses how scientists are punished for being popularizers. But it is a skill that can be acquired, not really more difficult than differential equations. In any case, if you won't spend the time, or don't want to be known as a popularizer, then simply turn down public debates. This is not an amateur's game. The stakes are way, way too high.

Messaging 101b: EcoAmerica's Phrase "Our Deteriorating Atmosphere" Isn't Going to Replace "Global Warming"—And That's a Good Thing

May 3, 2009

In a front-page article Saturday, "Seeking to Save the Planet, with a Thesaurus," the *New York Times* opens with some mostly bad messaging advice from ecoAmerica:

> The problem with global warming, some environmentalists believe, is "global warming."
>
> The term turns people off, fostering images of shaggy-haired liberals, economic sacrifice, and complex scientific disputes, according to extensive polling and focus group sessions conducted by ecoAmerica, a nonprofit environmental marketing and messaging firm in Washington.
>
> Instead of grim warnings about global warming, the firm advises, talk about "our deteriorating atmosphere." Drop discussions of carbon dioxide and bring up "moving away from the dirty fuels of the past." Don't confuse people with cap and trade; use terms like "cap and cash back" or "pollution reduction refund."

Yes, ecoAmerica is pushing the inapt phrase, "our deteriorating atmosphere" over "global warming" (and even over "climate change"). And ecoAmerica recommends generally skipping or dumbing down most of the climate science message. And ecoAmerica is pushing stuff that is just plain counterproductive—I quote now from material they handed out at a two-hour presentation I attended last week:

> It is also important to accept people's uncertainty about climate change but move past it with messages such as "whatever caused it, scientists know what will fix it."

Not. Definitely not. I'm not sure it even makes sense for moderate politicians talking to groups of swing voters and trying to push a very, very short-term message about the climate bill. But I'm quite certain it would be a suicidal message for climate science activists, for anyone seriously concerned about averting catastrophic global warming.

We know what is causing global warming and climate change. To suggest that we don't is the equivalent of undermining the essential credibility of our message and of what we are trying to do—help the public and policy makers make decisions based on science to preserve the health and well-being of their children, grandchildren, and the next fifty generations.

I have previously argued that phrases like "whatever caused it, scientists know what will fix it" are pure gobbledygook. If humans are not the cause of global warming then in fact scientists don't know how to fix it.

Let me run through some of the reasons why their climate messaging analysis is neither reliable nor strategic:

> 1) Other recent messaging and polling analysis contradicts it. I heard an extended presentation just last month from a different group with their polling, and they did not recommend shying away from the science. They simply suggested not making it more than half your message. That's certainly what I recommend. Indeed, other people in the audience with me at the ecoAmerica presentation made the same point that they had seen polling with different conclusions.
>
> 2) Many of ecoAmerica's findings are based on dial group responses. If somebody has a controlled study on whether a phrase that gets a positive response in a dial group is actually more persuasive or more memorable over the long-term than a phrase that gets a negative response, I'd love to see it. Obviously if you give people a dial to turn when you are telling them bad news—"you have diabetes"—they aren't going to like to hear that message. But if you are a credible source, I suspect they are more likely to take action (especially as more symptoms reveal themselves) than if you just tell them—"eating tasty fruits and crunchy vegetables will help you live longer," which would probably score much better on a dial group. You need both messages. Equally.
>
> In fact, I believe it was ecoAmerica's president and founder, Robert Perkowitz, who conceded to me after the meeting that people will sometimes give a negative dial response to a message that in fact turns out to

be an effective and persuasive messaging strategy. I believe that ecoAmerica work's on clean energy messaging is pretty good, as I discussed in Part 1, because I think the dial groups can tell you which positive message works better than another positive message. But I think the dial groups are largely useless for helping you with what I'd call "reality-based messaging" on climate.

3) The dial groups (but even the focus groups to a certain extent) are essentially passive ways to measure a response to a message to a very targeted audience. If you were giving a short speech in front of a large group of swing voters, then in that narrow case, their results might be useful to factor in. The vast majority of people I know don't do that. We sometimes go out and give talks, but then we have a lot of time to explain ourselves. I also have serious doubts about using some of their suggestions for many other audiences, including the media.

4) "Our deteriorating atmosphere" is a dead-end phrase. It is too inapt and unwieldy to be picked up by the key message pushers. ecoAmerica wants to tie in the general frame of "pollution threatens your health and your children's health." That is always worth talking about. But it is very hard to see how a six-syllable word is going to be a core element of successful messaging, especially in a passive phrase like that. Now ecoAmerica was also pushing the word "damage," and I do think an active phrase like "we are damaging the atmosphere," isn't bad. But the message isn't strategic, which brings me to the key point.

5) We are engaged in a multiyear messaging struggle here. The planet is going to get hotter and hotter, the weather is going to get more extreme. One of the reasons to be clear and blunt in your messaging about this is that even if you don't persuade people today, the overall message will grow in credibility as reality unfolds as we have warned. To shy away from telling people the truth because they don't want to hear it or they think it's liberal claptrap is just incredibly un-strategic. ecoAmerica doesn't want people to talk about "global warming." And—even worse—they don't want people to talk about extreme weather, which, as I have previously argued, is in fact the same thing that the climate deniers want. You *must* tell people what is coming, not just because it is strategic messaging, but also, I believe, because we have a moral responsibility.

True, if you use phrases like "global warming" you will activate certain frames in the audience because the right wing and fossil fuel disinformation machine have done a tremendous job politicizing this issue, making it seem like just another liberal-conservative argument, rather than a science versus anti-science argument. But does that mean we concede the powerful science frame just because the other side is more consistently effective and repetitious with their messaging? No.

The *New York Times* reports:

> Environmental issues consistently rate near the bottom of public worry, according to many public opinion polls. A Pew Research Center poll released in January found global warming last among 20 voter concerns; it trailed issues like addressing moral decline and decreasing the influence of lobbyists. "We know why it's lowest," said Mr. Perkowitz, a marketer of outdoor clothing and home furnishings before he started ecoAmerica, whose activities are financed by corporations, foundations and individuals. "When someone thinks of global warming, they think of a politicized, polarized argument. When you say 'global warming,' a certain group of Americans think that's a code word for progressive liberals, gay marriage, and other such issues."

I think Perkowitz has this backward. Why is it "lowest"? Lots of reasons. Probably the two most important factors that drive what the public thinks is important are (1) How the major news stories are framed by the media, and (2) What the White House focuses its messaging effort on. The media by and large downplay the issue—since they basically believe they "did global warming" back in 2006 with Gore's movie and in 2007 with the IPCC report. Also, environmentalists who talk to the media follow ecoAmerica's bad advice and largely fail to tell the public about the link between extreme weather and global warming.

The media also downplay the issue because their primary source for information on climate science—climate scientists—also downplay the issue. As one top UK environmental editor wrote recently, "Far from over-playing their hand to swell their research coffers, scientists have been toning down their message in an attempt to avoid public despair and inaction."

And let's please remember a 2007 report by the House Oversight and

Government Reform Committee concluded: "The Bush administration has engaged in a systematic effort to manipulate climate change science and mislead policy makers and the public about the dangers of global warming." For eight years! So I don't think it's a big surprise that global warming is not a bigger issue for the public, especially in the midst of the biggest recession since the Great Depression.

Let's remember that while progressive messaging has been scattershot at best, the right-wing disinformers have been both persistent and effective in their disinformation campaign. They have politicized this issue and pushed a partisan framing. But that is hardly a reason for climate science activists to give up explaining the issue to the public.

The answer, Mr. Perkowitz said in his presentation at the briefing, is to reframe the issue using different language. "Energy efficiency" makes people think of shivering in the dark. Instead, it is more effective to speak of "saving money for a more prosperous future." In fact, the group's surveys and focus groups found, it is time to drop the term "the environment" and talk about "the air we breathe, the water our children drink."

"Another key finding: remember to speak in TALKING POINTS aspirational language about shared American ideals, like freedom, prosperity, independence, and self-sufficiency while avoiding jargon and details about policy, science, economics or technology," said the e-mail account of the group's study.

Well, I'm all for dropping the word "environment" (see my Earth Day piece below). As I've said, Messaging 101 is to be specific. Yes, "jargon" is bad. But details about science, economics, and technology are what people are hungry for. They are what make you seem more credible.

Yes, if you are giving a 10-minute speech in front of swing voters, skip the details. Duh. But most of the rest of the time this just isn't good advice.

Yes, aspirational language is important to use. But a core tenet of rhetoric is to speak truthfully about what you know. If you don't know the climate science, then you probably shouldn't talk about it. But frankly if you don't know the science, you will be eaten alive by the informed conservative doubters in your audience, not to mention any professional deniers you might be debating or who might be on the same panel. A classic technique of rhetoric and debating is to go after your opponent on whatever they are weakest. That's why you need to know the science and how to explain it and defend it.

So if you are out there pushing gobbledygook, a savvy conservative or clever contrarian (or even a sharp reporter) will make you look like an uninformed fool. Remember the key line of the smarmy tobacco lobbyist in the must-see movie, *Thank You for Smoking*: "I don't have to be right. I just have to prove you might be wrong."

So the anti-science disinformers have the easier end of it on global warming messaging—they can throw out 100 lies and succeed if even one sticks. That's no reason to walk away from the science. Quite the reverse.

Indeed, I think at some level, some of ecoAmerica's recommendations are elitist, suggesting that we can't explain the facts to swing voters—that they can't handle the truth—but instead we need to use obscuring or vague phrases to persuade them. I couldn't disagree more. And I'm not alone.

The fact that our best communicator—President Obama, "The FDR of clean, safe sources of energy that never run out"—takes every opportunity he has to speak about capping carbon dioxide and avoiding catastrophic global warming impacts is probably the clearest evidence that the rest of us climate messaging amateurs should also keep doing so.

I mostly agree with this sentiment from the end of the *New York Times* article:

> Robert J. Brulle of Drexel University, an expert on environmental communications, said ecoAmerica's campaign was a mirror image of what industry and political conservatives were doing. "The form is the same; the message is just flipped," he said. "You want to sell toothpaste, we'll sell it. You want to sell global warming, we'll sell that. It's the use of advertising techniques to manipulate public opinion."
>
> He said the approach was cynical and, worse, ineffective. "The right uses it, the left uses it, but it doesn't engage people in a face-to-face manner," he said, "and that's the only way to achieve real, lasting social change."

I do fully agree that engaging people in a face-to-face manner with the truth is the only way to achieve real, lasting social change.

But I don't think there's anything wrong with using the best techniques of rhetoric (even if they have been partly rediscovered and abused by the advertising industry) in your messaging. You'll never win a debate against a skilled debater without rhetoric. But you'll never win on this issue by downplaying the scientific reality.

What you need to know is *not* how to avoid talking about climate science to the public. What you need to know is *how* to talk about climate science to the public—and that is a key focus of this blog.

Let's Dump "Earth Day"

April 19, 2009

Affection for our planet is misdirected and unrequited. We need to focus on saving ourselves.

Last year, I wrote a piece for *Salon*, "Let's Dump 'Earth' Day." It was supposed to be mostly humorous. Or mostly serious. Anyway, the subject of renaming Earth Day has been on my mind for a year now—and all the more so today because the *New York Times* magazine just published an interview with our Nobel Prize–winning energy secretary, Steven Chu, in which he says "I would say that from here on in, every day has to be Earth Day."

Well, duh! Heck, we have a whole day just for the trees—and we haven't finished them off . . . yet. So if every day is Earth Day, then April 22 definitely needs a new name. So I'm updating the column, with yet another idea at the end, at least for climate science advocates.

I don't worry about the Earth. I'm pretty certain the Earth will survive the worst we can do to it. I'm very certain the Earth doesn't worry about us. I'm not alone. People got more riled up when scientists removed Pluto from the list of planets than they do when scientists warn that our greenhouse gas emissions are poised to turn the Earth into a barely habitable planet.

Arguably, concern over the Earth is elitist, something people can afford to spend their time on when every other need is met. But elitism is out these days. Only environmentalists cling to Earth Day. We need a new way to make people care about the nasty things we're doing with our cars and power plants. At the very least, we need a new name.

How about Nature Day or Environment Day? Personally, I am not an environmentalist. I don't think I'm ever going to see the Arctic National Wildlife Refuge. I wouldn't drill for oil there. But that's not out of concern for the caribou but for my daughter and the planet's next several billion people, who will need to see oil use cut sharply to avoid the worst of climate change.

I used to worry about the polar bear. But then some naturalists told me that once human-caused global warming has destroyed their feeding habitat—the polar ice, probably by 2020, possibly sooner—polar bears will just go about the business of coming inland and attacking humans and eating our food and maybe even us. That seems only fair, no?

I am a cat lover, but you can't really worry about them. Cats are survivors. Remember the movie *Alien*? For better or worse, cats have hitched their future to humans, and while we seem poised to wipe out half the species on the planet, cats will do just fine.

Apparently there are some plankton that thrive on an acidic environment, so it doesn't look like we're going to wipe out all life in the ocean, just most of it. Sure, losing Pacific salmon is going to be a bummer, but I eat Pacific salmon several times a week, so I don't see how I'm in a position to march on the nation's capital to protest their extinction. I won't eat farm-raised salmon, though, since my doctor says I get enough antibiotics from the tap water.

If thousands of inedible species can't adapt to our monomaniacal quest to return every last bit of fossil carbon back into the atmosphere, why should we care? Other species will do just fine, like kudzu, cactus, cockroaches, rats, scorpions, the bark beetle, Anopheles mosquitoes, and the malaria parasites they harbor. Who are we to pick favorites?

I didn't hear any complaining after the dinosaurs and many other species were wiped out when an asteroid hit the Earth and made room for mammals and, eventually, us. If God hadn't wanted us to dominate all living creatures on the Earth, he wouldn't have sent that asteroid in the first place, and he wouldn't have turned the dead plants and animals into fossil carbon that could power our Industrial Revolution, destroy the climate, and ultimately kill more plants and animals.

And speaking of God, Creation Care is also woefully misnamed. If humans are special, invested with a soul by our Creator, along with the right to life, liberty, and the pursuit of happiness, then why should we sacrifice even a minute of that pursuit worrying about the inferior species?

All of these phrases create the misleading perception that the cause so many of us are fighting for—sharp cuts in greenhouse gases—is based on the desire to preserve something inhuman or abstract or far away. But I have to say that all the environmentalists I know—and I tend to hang out with the climate crowd—care about stopping global warming because of its impact on humans, even if they aren't so good at articulating that perspective. I'm with them.

The reason that many environmentalists fight to save the Arctic National Wildlife Refuge or the polar bears is not because they are sure that losing those things would cause the universe to become unhinged, but because they realize that humanity isn't smart enough to know which things are linchpins for the entire ecosystem and which are not. What is the straw that breaks the camel's back? The hundredth species we wipe out? The thousandth? For many, the safest and wisest thing to do is to try to avoid the risks entirely.

This is where I part company with many environmentalists. With 6.5 billion people going to 9 billion, much of the environment is unsavable. But if we warm significantly more than 4°F from preindustrial levels—and especially if we warm more than 7°F, as would be all but inevitable if we keep on our current emissions path for much longer—then the environment and climate that made modern human civilization possible will be ruined, probably for hundreds of years. And that means misery for many if not most of the next 10 to 20 billion people to walk the planet.

So I think the world should be more into conserving the stuff that we can't live without. In that regard I am a conservative person. Unfortunately, Conservative Day would, I think, draw the wrong crowds.

The problem with Earth Day is it asks us to save too much ground. We need to focus. The two parts of the planet worth fighting to preserve are the soils and the glaciers.

Two years ago, *Science* magazine published research that "predicted a permanent drought by 2050 throughout the Southwest"—levels of soil aridity comparable to the 1930s Dust Bowl would stretch from Kansas and Oklahoma to California. The Hadley Centre, the UK's official center for climate change research, found that "areas affected by severe drought could see a five-fold increase from 8 percent to 40 percent." On our current emissions path, most of the South and Southwest will ultimately experience twice as much loss of soil moisture as was seen during the Dust Bowl.

Also, locked away in the frozen soil of the tundra or permafrost is more carbon than the atmosphere contains today. On our current path, most of the top 10 feet of the permafrost will be lost this century—so much for being "perma"—and that amplifying carbon-cycle feedback will all but ensure that today's worst-case scenarios for global warming will become the best-case scenarios. We must save the tundra. Perhaps it should be small "e" earth Day, which is to say, Soil Day. On the other hand, most of the public enthusiasm in the 1980s for saving the rain forests fizzled, and they are almost as important as the soil, so maybe not Soil Day.

As for glaciers, when they disappear, sea levels rise, perhaps as much as 2 inches a year by century's end. (A 2009 *Nature* paper on sea level rise using data from coral fossils suggested "catastrophic increase of more than 5 centimetres per year over a 50-year stretch is possible." The lead author warned, "This could happen again."). If we warm even 7°F from preindustrial levels, we will return the planet to a time when sea levels were ultimately 80 feet higher. The first 5 feet of sea level rise, which seems increasingly likely to occur this century on our current emissions path, would displace more than 100 million people. That would be the equivalent of 200 Katrinas. Since my brother lost his home in Katrina, I don't consider this to be an abstract issue.

Equally important, the inland glaciers provide fresh water sources for more than a billion people. But on our current path, most of them will be gone by century's end.

So where is everyone going to live? Hundreds of millions will flee the new deserts, but they can't go to the coasts; indeed, hundreds of millions of other people will be moving inland. But many of the world's great rivers will be drying up at the same time, forcing massive conflict among yet another group of hundreds of millions of people. The word "rival," after all, comes from "people who share the same river." Sure, desalination is possible, but that's expensive and uses a lot of energy, which means we'll need even more carbon-free power.

Perhaps Earth Day should be Water Day, since the worst global warming impacts are going to be about water—too much in some places, too little in other places, too acidified in the oceans for most life. But even soil and water are themselves only important because they sustain life. We could do Pro-Life Day, but that term is already taken, and again it would probably draw the wrong crowd.

We could call it *Homo sapiens* Day. Technically, we are the subspecies *Homo sapiens sapiens.* Isn't it great being the only species that gets to name all the species, so we can call ourselves "wise" twice! But given how we have been destroying the planet's livability, I think at the very least we should drop one of the sapiens. And, perhaps provisionally, we should put the other one in quotes, so we are Homo "sapiens," at least until we see whether we are smart enough to save ourselves from self-destruction.

What the day—indeed, the whole year—should be about is not creating misery upon misery for our children and their children and their children, and on and on for generations. Ultimately, stopping climate change is not about preserving the Earth or creation but about preserving ourselves. Yes, we can't preserve ourselves if we don't preserve a livable climate, and we can't preserve a livable climate if we don't preserve the Earth. But the focus needs to stay on the health and well-being of billions of humans because, ultimately, humans are the ones who will experience the most prolonged suffering. And if enough people come to see it that way, we have a chance of avoiding the worst.

We have fiddled like Nero for far too long to save the whole Earth or all of its species. Now we need a World War II–scale effort just to cut our losses and save what matters most. So let's call it Triage Day. And if worse comes to worst—yes, if worse comes to worst—at least future generations won't have to change the name again.

Murder, He Wrote

March 1, 2009

An excellent climate blogger, Johnny Rook, is dying. I'm hoping that you will take a look at his website, Johnny Rook's Climaticide Chronicles, and post a comment for his family.

His most recent post, no doubt a great struggle to write, has some breaking news: "Alert: Wilkins Ice Shelf Collapses According to Spanish Scientists." As you can see he hasn't been able to post much recently for reasons he explains in his February 5 post "My Doctor Doesn't Think I'm Going to Die Today—Updated."

Rook makes a plea in that post I'll repeat here because it is something that you perhaps can respond to in your comments:

> My diaries, as those of you who are regular readers know, often contain depressing information about how temperatures and sea levels are rising, how sea ice and glaciers are melting and shrinking, how deserts are growing and heat waves becoming longer and hotter meaning that agriculture is becoming less and less possible in many places, how extreme weather is becoming more common and more intense, how oceans are becoming acidified, how species are going extinct, ecosystems are being rendered uninhabitable for the creatures that live in them, and how famine and diseases are spreading.
>
> When I write about solutions I often focus on how people and governments are mostly oblivious to what is happening and to how little time we have left to act boldly and forcefully to effect the radical change that the scientists tell us is necessary. I agree absolutely with what Steven Chu, the new secretary of energy, told the *LA Times* in an interview a couple of days ago.
>
> I don't think the American public has gripped in its gut what could happen.

> I understand that such news can depress. At times it depresses me but, more than anything else, it has filled my life with meaning. I have a mission. Before I die, I want to have some sense that this beautiful planet that has provided the context for my life will have some chance of enduring. I want to die with hope, believing that my teenage son and his children and your children and their children will live in a world that is reasonably hospitable to human beings.
>
> I don't know how that can happen if people will not face the reality of what is taking place in the world. So, I continue to sound the alarm, even though I know that most of what I write is discounted as alarmist or simply ignored as too uncomfortable to deal with.

Johnny Rook shares my initials and my sensibilities on climate, so I understand how optimism can be hard to come by.

My father, while he was alive, watched *Murder, She Wrote*—the TV series that starred Angela Lansbury as mystery writer and amateur detective Jessica Fletcher. He would joke that the most dangerous place on Earth was in the company of Jessica Fletcher, because wherever she was, somebody was going to get murdered.

Humanity is, as the name Climaticide Chronicles makes clear, in the process of murdering the climate. Everywhere Rook—or any of us—looks, there is more and more evidence of that crime in process.

But it is not too late. The murder can be stopped. I wouldn't be blogging if didn't know that for a fact. The fact that catastrophic climate change, say, post-2040, is irreversible does not mean it is unstoppable if we act now. I'm sure Rook feels the same way.

So there is hope as long as people like Johnny Rook are willing to use their energy—even their last drop of energy—to tell the world what is to come on our current path and how we can stop it.

I hope you'll tell his family that.

> *Johnny Rook* Says:
> March 1st, 2009 at 4:56 pm
> Dear Joe,
> Thank you for such a touching encomium.

> Humanity is, as the name Climaticide Chronicles makes clear, in the process of murdering the climate. Everywhere Rook—or any of us—looks, there is more and more evidence of that crime in process.
>
> But it is not too late. The murder can be stopped. I wouldn't be blogging if didn't know that for a fact. . . . I'm sure Rook feels the same way.

[I do-Rook]

I am proud to say that my nineteen-year-old son, Aleks and a friend, Andrew, did make the choice to go in my stead and are already in DC for the Powershift Action. I know this is going to be a very exciting time for both of them, and that they will carry the battle into the future until we have succeeded.

It's getting harder and harder for me to write, even simple comments like this, but my thoughts and sympathies are with you always.

UPDATE: Johnny Rook died the next day.

CONCLUSION

Is the Global Economy a Ponzi Scheme?

March 8, 2009

YES, HOMO "SAPIENS" have constructed the grandest of Ponzi schemes, whereby current generations have figured out how to live off the wealth of future generations. Yes, we are all in essence Madoffs (a few wittingly, most not), or at least his most credulous clients. What comes next will be the subject of a multipart series.

I had been planning to write something on this for a while when *New York Times* columnist Tom Friedman interviewed me for "The Inflection Is Near?" which appears in today's *Times*:

> "We created a way of raising standards of living that we can't possibly pass on to our children," said Joe Romm, a physicist and climate expert who writes the indispensable blog ClimateProgress.org. We have been getting rich by depleting all our natural stocks—water, hydrocarbons, forests, rivers, fish, and arable land—and not by generating renewable flows.
>
> "You can get this burst of wealth that we have created from this rapacious behavior," added Romm. "But it has to collapse, unless adults stand up and say, 'This is a Ponzi scheme. We have not generated real wealth, and we are destroying a livable climate. . . .' Real wealth is something you can pass on in a way that others can enjoy."

A few years ago I thought that aggressive action by governments around the world to push clean energy could spare the public dramatic lifestyle changes in the coming decades, but I have been convinced otherwise by:

- The failure of U.S. leadership (thank you, George W. Bush and the conservative stagnation);
- The remarkable shift in our understanding of climate science during the last two years;
- China's decision to join the Ponzi scheme full throttle and emulate our rapaciousness; and
- A recent, brilliant talk I heard by Saul Griffiths.

The adults, in short, are not standing up. Sadly, most, like pundit David Broder, haven't even taken the time to understand that they should.

And so every generation that comes after the Baby Boomers is poised to experience the dramatic changes in lifestyle that inevitably follow the collapse of any Ponzi scheme. This global Ponzi scheme is not just a metaphor, but for me a central organizing narrative of how to think about the fix we have put ourselves in.

What exactly is a Ponzi scheme? Wikipedia makes several good points:

> A Ponzi scheme is a fraudulent investment operation that pays returns to investors from their own money or money paid by subsequent investors rather than from profit. The term "Ponzi scheme" is used primarily in the United States, while other English-speaking countries do not distinguish colloquially between this scheme and pyramid schemes.
>
> The Ponzi scheme usually offers abnormally high short-term returns in order to entice new investors. The perpetuation of the high returns that a Ponzi scheme advertises and pays requires an ever-increasing flow of money from investors in order to keep the scheme going.

In our case, investors (i.e., current generations) are paying themselves (i.e., you and me) by taking the nonrenewable resources and livable climate from future generations. To perpetuate the high returns the rich countries in particular have been achieving in recent decades, we have been taking an ever greater fraction of nonrenewable energy resources (especially hydrocarbons) and natural capital (fresh water, arable land, forests, fisheries), and the most important nonrenewable natural capital of all—a livable climate.

> The system is destined to collapse because the earnings, if any, are less than the payments.

Same for ours—thanks to climate change and peak oil.

> Usually, the scheme is interrupted by legal authorities before it collapses because a Ponzi scheme is suspected or because the promoter is selling unregistered securities.

Yes, well, the authorities (i.e., world leaders, opinion makers, the cognoscenti) haven't been doing much interrupting over the last two to three decades since, unlike a typical Ponzi scheme, they are heavily invested in the scheme and addicted to the returns!

> As more investors become involved, the likelihood of the scheme coming to the attention of authorities increases.

Well now I do think that the scheme has come to the attention of many of "the authorities," at least to many leaders around the world and to progressive ones here at home. Conservative authorities simply have too much invested in the status quo.

> Knowingly entering a Ponzi scheme, even at the last round of the scheme, can be rational in the economic sense if a government will likely bail out those participating in the Ponzi scheme.

But Friedman quotes Glenn Prickett, senior vice president at Conservation International, explaining, "Mother Nature doesn't do bailouts."

We aren't all Madoffs in the sense of people who have knowingly created a fraudulent Ponzi scheme for humanity. But given all of the warnings from scientists and international governments during the last quarter-century (most recently two years ago with "Absolute MUST Read IPCC Report: Debate Over, Further Delay Fatal, Action Not Costly"; see chapter 2)—it has gotten harder and harder for any of us to pretend that we are innocent victims, that we aren't just hoping we can maintain our own personal wealth and well-being for a few more decades before the day of reckoning.

But it is interesting to see just how clever Madoff was:

> Madoff's scheme was typical of a Ponzi in its structure, but differed in its pace and marketing. Rather than offer (suspiciously) high returns to all comers, Madoff offered modest, but steady returns to an exclusive clientele, produced in both up and down markets. Although the investment method was marketed as a "too complicated for outsiders to

> understand" . . . the true secret to Madoff's success was his lifetime involvement with nonprofit charities and the tax law knowledge he gleaned from that experience over many decades.
>
> Charitable foundations were the basis, as well as the side-victims, of his surreptitious strategy. . . . The slow pace and ongoing cliquish "insider" word-of-mouth marketing enabled the deception to survive for several decades. It grew beyond the expectations of a common Ponzi. . . .
>
> Mitchell Zuckoff, professor of journalism at Boston University and author of *Ponzi's Scheme: The True Story of a Financial Legend*, explained, "By targeting charities, Madoff could avoid the threat of sudden or unexpected withdrawals." Zuckoff suggests that years ago, Madoff "solved the two interlocking puzzles that usually prevent Ponzi schemes from becoming perpetual money machines: sustaining growth, while maintaining stability."

But humanity has made Madoff look like a penny-ante criminal.

By enriching the authorities, as noted, we encouraged those who have the most power to solve the problem to do nothing.

By enriching those who did the most plundering, we enabled them to fund lobbying and disinformation campaigns to convince substantial fractions of the public and media that there is no Ponzi scheme—that global warming is "too complicated for the public to understand" and nothing to worry about.

And by "paying ourselves" with the wealth from future generations—indeed, from the next fifty generations and next 100 billion people to walk the Earth—we cleverly took advantage of victims not yet born, those not able to even know they were being robbed.

Madoff is reviled for targeting charities. We are targeting our own children and grandchildren and on and on. What does that make us?

Afterword

The Copenhagen Climate Conference Made Clear a Global Deal Is Within Reach—If the United States Can Act

January 3, 2010

We've just left the hottest decade on record, just as we did 10 years ago when the 1990s ended, according to NASA. We've just entered what will doubtless also be the hottest decade on record, much as we will 10 years from now when the 2020s start.

This year, which marks the 40th anniversary of Earth Day on April 22, may well determine whether every decade this century will become the hottest decade on record—taking us up to 10°F planetary warming by the 2090s, accompanied by catastrophic sea-level rise, widespread dust-bowlification, an acidified ocean, and the destruction of a livable climate—or whether the nation and the world are wise enough to reverse our greenhouse gas (GHG) emissions trend quickly and sharply.

After spending a week in Copenhagen during the huge global climate negotiations in December 2009, and talking to people from around the world as well as leading administration officials and members of Congress, I think it is now clear that virtually every major emitting country in the world is prepared to take strong action—if the United States will act.

Yes, I know that many in the media declared Copenhagen a failure before it even began (see my October 2009 post, "*NY Times* Spins the Greatest Nonstory Ever Told, Suckering UK's *Guardian* into Printing Utter BS").

Googling "Copenhagen" and "failure" gets you 9 million hits! But those who follow international negotiations know that the Copenhagen Summit achieved a great deal, as we'll see.

I have not, however, been fond of how the United Nations has been running all things climate. Both my colleague Andrew Light of the Center for American Progress and I have argued before that "we don't need 193 nations to come to an agreement on mitigating carbon emissions in order to get the job done. We only need those countries responsible for 85% of emissions to move forward on the pathways identified by the IPCC with a promise to the world to do so in a responsible manner."

That's why much of what 350.org founder (and occasional ClimateProgress guest blogger) Bill McKibben doesn't like about the Copenhagen Accord is exactly what I like about it. McKibben complains of Obama's successful effort to prevent a complete failure at Copenhagen:

> He blew up the United Nations. . . .
> He formed a league of super-polluters, and would-be super-polluters. . . .

Hurray!

Why is this a good thing? Nobelist Paul Krugman wrote of the Congressional debate over health care last month, "the fact that it was such a close thing shows that the Senate—and, therefore, the U.S. government as a whole—has become ominously dysfunctional." And yet this "dangerous dysfunction," as he puts it, is due solely to the need for a 60% supermajority:

> . . . the need for 60 votes to cut off Senate debate and end a filibuster—a requirement that appears nowhere in the Constitution, but is simply a self-imposed rule—turned what should have been a straightforward piece of legislating into a nail-biter. And it gave a handful of wavering senators extraordinary power to shape the bill.

Now imagine how much the United States would accomplish if every single member of Congress had a veto! Well, that's the UN Framework Convention on Climate Change (UNFCCC) process. So that process needed to be changed significantly or ended entirely. Kudos to Secretary of State Hillary Clinton and President Barack Obama for realizing that and working to bring it about—working to make sure that the biggest GHG emitters achieved many of the key elements of a global deal—even if their real objectives weren't widely understood and even if it meant sacrificing the possibility that Copenhagen achieved a unanimous and binding deal.

Ironically, for those who want to avoid total planetary warming of 2°C (3.6°F) or higher—as McKibben does—it was China who was a bigger obstacle than America in the final days at Copenhagen. Still clinging to the Kyoto approach for most of the two weeks of negotiations, where developing countries don't have to commit to anything, China also almost single-handedly made it impossible for anyone modeling the commitments to report that the world could come anywhere near those targets. So this allowed the media and others to assert that Copenhagen wouldn't achieve the 2°C target if you just added up all of the nations' commitments, as if that actually meant the conference was a failure or worse.

China, and to a lesser extent India, had been hiding behind U.S. (i.e., Bush–Cheney) inaction for 8 years. And as long as we kept the Kyoto Protocol process, they could hide behind the Sudans and Ethiopias of the world indefinitely. With Obama providing as much leadership as possible given our dysfunctional Senate, and with Clinton coming to Denmark with a real commitment to work with the other rich countries to deliver tens of billions of dollars a year to help poor countries develop with low-carbon energy, China was left in the role of spoiler (see my blog post, "Clinton's $100-Billion Copenhagen Bombshell"). And that's essentially what they tried to do, as the UK's Climate Change and Energy Secretary, Ed Miliband, wrote in *The Guardian*:

> We did not get an agreement on 50% reductions in global emissions by 2050 or on 80% reductions by developed countries. *Both were vetoed by China,* despite the support of a coalition of developed and the vast majority of developing countries. Indeed, this is one of the straws in the wind for the future: the old order of developed versus developing has been replaced by more interesting alliances.

Specifically, they were vetoed by China's lead climate negotiator. But the rich countries decided not to walk away from talks. President Obama personally intervened to cut a deal in a key smaller meeting with the heads of the big developing country emitters: Indian Prime Minister Manmohan Singh, Brazilian President Luiz Inácio Lula da Silva, South African President Jacob Zuma, and Chinese premier Wen Jiabao.

As Robert Orr, the UN assistant secretary-general for policy, and deputy to our UN ambassador during the Clinton administration, told *Greenwire*:

> This was the most genuine negotiation I've ever seen between leaders. I've never seen leaders truly negotiate. It's usually prearranged. It's precooked. And the text goes to the leaders, they nod at each other and they agree. This was not the case. Leaders were drafting. Leaders were caucusing. All of them were doing things that most of them had probably not done for a few years. And for many of them, they even said, I haven't done this for awhile.

Miliband himself noted the significance of what was achieved:

> Countries signing the accord have endorsed the science that says we must prevent warming of more than 2°C. For the first time developing countries, including China, as well as developed countries have agreed [to] emissions commitments for the next decade. If countries deliver on the most ambitious targets, we will be within striking distance of what is needed to prevent warming of more than 2°C. These commitments will also for the first time be listed and independently scrutinised, with reports to the UN required every two years.

Now it is true that before Copenhagen, China put out a meaningful carbon intensity goal that takes them significantly off of the business-as-usual emissions path (see my blog post, "China Vows to Dramatically Slow Emissions Growth"). But in fact they are going to have to do more if we are to have any chance at 2°C (and yes, America will need to do more, too):

- China will need to beat their goal of cutting carbon intensity 40% to 45% from 2005 to 2020.
- They'll need to peak in emissions around 2020–2025 and then reduce emissions steadily after that.

And I'm quite certain the Chinese know that and that they will do both. Indeed, I had a discussion with a number of senior administration officials right after Copenhagen ended, and they confirmed that China's leaders know they must do both of those, and also that they know they are easily going to beat their 2020 carbon intensity target.

And no, I don't entirely blame China here, since they moved a great deal in the final 36 hours (albeit not far enough), as the National Wildlife Federation's Jeremy Symons explained:

> Most importantly, China is now officially in the game in a way it has resisted since the Earth Summit almost two decades ago. The Copenhagen Accord is a two-part breakthrough with China: They are putting numbers on the table with a measurable pledge to join the global fight to reduce climate pollution, and they agreed to open their books on their rising emissions and allow a transparent review of their progress toward their emissions pledge. This breakthrough is important for the global climate effort, as well as encouraging the Senate to move forward and deliver the climate and clean energy bill to the president. China will act, and the China excuse is off the table.

I have long said that achieving serious global action climate required Congress to pass bipartisan legislation, which in turn required a bilateral deal with China. Well, it seems to me that team Obama's efforts in this area are paying off (see "Have China and the U.S. Been Holding Secret Talks Aimed at a Climate Deal This Fall?" in chapter 5). One couldn't have realistically expected a full deal with the key emitters in the context of the 193-country UNFCCC meeting in Copenhagen.

I agree with Reuters that Copenhagen "underscored the vulnerability of a process depending on consensus and may mark a diminishing UN role." Ultimately, the point is not the friggin' process, but the outcome, and if the UN could demonstrate its process could lead to a better outcome, I'd be all for it. But I doubt it.

I think Obama showed the process that can work to get the best possible outcome: High-level negotiations by the senior leaders of the big emitters.

I also agree with the conclusion of Harvard economist Robert Stavins:

> We may look back upon Copenhagen as an important moment—both because global leaders took the reins of the procedures and brought the

> negotiations to a fruitful conclusion, and because the foundation was laid for a broad-based coalition of the willing to address effectively the threat of global climate change. Only time will tell.

For me, the Copenhagen "glass" is 2/3 full, since the point wasn't just the meeting, but the remarkable commitments that countries made leading up to the meeting. As CAP's Andrew Light explained in Copenhagen:

> When you add up everything that the 17 largest economies have on the table, not for a treaty mind you, but awaiting domestic action that could happen regardless of a treaty, such as the U.S. legislation, then we are 5 gigatons away from commitments that should get us on a 450 ppm stabilization path by 2020, essentially 65% of the way there. Given that the world has managed to get on a potential track in that direction with the world's largest historical emitter pretending nothing was happening in the mean time, and only trying to catch up recently, isn't bad at all.

Moreover, what happens after 2020 is probably even more important, and here the United States is on the verge of making a true leadership commitment, if the Senate passes the bipartisan climate and clean energy bill. And if we do, then I expect that should be enough to get China and the other big emitters to formalize a binding deal.

Will Anti-Science Ideologues Be Able to Kill Domestic Climate and Clean Energy Action?

January 10, 2010

The fate of the international deal rests to a large extent on the fate of the bipartisan U.S. climate and clean energy bill, which passed the House in June 2009 and is too-slowly winding its way through the Senate. The bill continues to have broad public support, despite a massive disinformation campaign against it launched by the big polluters:

- Swing state poll finds 60% "would be more likely to vote for their senator if he or she supported the bill" and Independents support the bill 2-to-1"
- New CNN poll finds "nearly six in 10 independents" support cap-and-trade
- Voters in Ohio, Michigan and Missouri overwhelmingly support action on clean energy and global warming
- Public Opinion Stunner: WashPost-ABC Poll Finds Strong Support for Global Warming Reductions Despite Relentless Big Oil and Anti-Science Attacks

One of the unique features of the House climate and clean energy bill is just how bipartisan it was. The final vote at the end of June 2009 (219-212) included a whopping eight Republicans. That may not seem like a lot, but in fact it is remarkable.

Consider that the economic stimulus bill passed during the depths of the worst recession in 70 years—to address a problem that was transparently urgent and far more politically salient—garnered precisely zero Republicans. Health care reform—one of the paramount issues of our time—garnered one House GOP vote. Financial services reform—a seeming political winner—again got zero GOP votes.

In the Senate, the stimulus bill got three GOP votes, one of which is now Democratic and the other two of which are from Maine. Healthcare reform received zero GOP votes when it passed the Senate on Christmas Eve 2009. But Senate climate action was long pursued by one Republican (McCain), and later advanced in 2008 by another conservative Republican (former Senator John Warner of Virginia). The current climate bill is now being actively pursued by a bipartisan team led by one of the more conservative members of the GOP, Lindsey Graham of South Carolina, in part because he is troubled that Americans spend nearly $1 billion a day "to sustain our addiction to foreign energy sources . . . some of which finds its way to extremist or terrorist organizations." Graham has further said:

> I believe the green economy is coming. That's not a question of if it's going to happen, it's just when it's going to happen. The sooner the better for me, because the jobs of the future lie in energy independence and cleaning up the environment.... Why can't America have the cleanest air?

UPDATE (1/21/10): The election of Republican Scott Brown to fill the Massachussetts Senate seat long held by Democrat Ted Kennedy means it is more true than ever that the fate of U.S. climate action relies on bipartisan support, relies on Republicans who are willing to stand up for clean air, clean energy jobs, and a livable climate.

Of course, outside of Washington, action on global warming has been exceedingly bipartisan. The governor who has done the most on climate action is, arguably, California's Arnold Schwarzenegger, a Republican. Other leaders have included current and former Republican governors like New York's George Pataki, Florida's Charlie Crist, and Minnesota's Tim Pawlenty.

If not for the Senate's "need" for 60 votes to pass almost any piece of legislation, the Senate would probably have passed its own climate bill in 2009. One key lesson of Copenhagen is that Obama is so committed to climate action that he is willing to negotiate personally with leaders to get a deal done, which is precisely what it will take to move the key fence-sitting senators of both parties.

Really, only one political force could stop a climate bill, the same force that has impeded action for more than a decade—the hard-core anti-science crowd that dominates much of conservative politics these days and that demagogues against even the most modest efforts to promote clean energy and reduce pollution. Indeed, they are pushing hard to punish any Republican who contemplates actually addressing the gravest threat to the health and well-being of our children and grandchildren. For instance, in September 2009, Republican National Committee Chairman Michael Steele withdrew his endorsement for Rep. Mark Kirk in the Illinois Senate race in large part because he was one of the eight House Republicans to vote for the House climate bill (see my blog post, "Honey, I Shrunk the GOP, Part 3").

Since Crist faces a right-wing challenger in the Florida Senate primary, he has moved away from his strong support for climate action. Pawlenty still dreams of being on the national ticket in 2012, so he too has abandoned his previous support for strong climate action.

This emerging conservative litmus test is in many respects unique to U.S. politics. As I noted in a December 2009 blog post on the British reaction to the stolen emails ("British PM Gordon Brown Attacks 'Anti-Science, Flat-Earth Climate Sceptics" While UK Conservatives Reaffirm Climate Science), the top environmental leader for the conservatives in Parliament understands both the science and the urgent need for action:

> But tonight the shadow climate change secretary, Greg Clark, made clear the party line remains that climate change is a serious man-made threat. "Research into climate change has involved thousands of different scientists, pursuing many separate lines of independent inquiry over many years. The case for a global deal is still strong and in many aspects, such as the daily destruction of the Earth's rainforests, desperately urgent," he said.

Yes, there is nothing genuinely "conservative" about refusing to conserve resources or refusing to conserve a livable climate.

Some claim futuristic "geo-engineering" strategies could avoid the need to reduce emissions quickly and sharply. They suggest we could cool the planet by putting millions of mirrors in space to reflect sunlight or inject massive amounts of particulates (sulfate aerosols) to partially block the sun. But those solutions are, literally, smoke and mirrors. They won't stop the oceans from being acidified, devastating corals and other marine life. Even those scientists who advocate geo-engineering research, like climatologist Ken Caldeira, have explained that it can't work unless the world first achieves very deep emissions reductions (see my blog post "Caldeira calls the vision of Lomborg's Climate Consensus 'a dystopic world out of a science fiction story'").

The science makes clear that no quick fixes exist to preserve a livable climate. The only solution is to start cutting global warming pollution as fast as possible using every available strategy. And that's why the fight in 2010 and beyond is not Democrat versus Republican or progressive versus conservative. It is science versus anti-science. Science will eventually prevail. It always does.

Index

C

D

E

F

G

H

About Island Press

Since 1984, the nonprofit Island Press has been stimulating, shaping, and communicating the ideas that are essential for solving environmental problems worldwide. With more than 800 titles in print and some 40 new releases each year, we are the nation's leading publisher on environmental issues. We identify innovative thinkers and emerging trends in the environmental field. We work with world-renowned experts and authors to develop cross-disciplinary solutions to environmental challenges.

Island Press designs and implements coordinated book publication campaigns in order to communicate our critical messages in print, in person, and online using the latest technologies, programs, and the media. Our goal: to reach targeted audiences—scientists, policymakers, environmental advocates, the media, and concerned citizens—who can and will take action to protect the plants and animals that enrich our world, the ecosystems we need to survive, the water we drink, and the air we breathe.

Island Press gratefully acknowledges the support of its work by the Agua Fund, Inc., The Margaret A. Cargill Foundation, Betsy and Jesse Fink Foundation, The William and Flora Hewlett Foundation, The Kresge Foundation, The Forrest and Frances Lattner Foundation, The Andrew W. Mellon Foundation, The Curtis and Edith Munson Foundation, The Overbrook Foundation, The David and Lucile Packard Foundation, The Summit Foundation, Trust for Architectural Easements, The Winslow Foundation, and other generous donors.